Toxicity of Pesticides to Fish

Volume II

Author

A. S. Murty
Department of Zoology
Nagarjuna University
Nagarjunanagar, India

CRC Press
Taylor & Francis Group
Boca Raton London New York

CRC Press is an imprint of the
Taylor & Francis Group, an **informa** business

First published 1986 by CRC Press
Taylor & Francis Group
6000 Broken Sound Parkway NW, Suite 300
Boca Raton, FL 33487-2742

Reissued 2018 by CRC Press

Library of Congress Cataloging in Publication Data

Murty, A. S., 1943—
    Toxicity of pesticides to fish.

    Bibliography: p.
    Includes index.
    1. Fishes—Effect of water pollution on.
2. Pesticides—Toxicology. I. Title.
SH174.M87   1986   597'.024   85-4182
ISBN 0-8493-6058-7 (v. 1)
ISBN 0-8493-6059-5 (v. 2)

A Library of Congress record exists under LC control number: 85004182

Publisher's Note
The publisher has gone to great lengths to ensure the quality of this reprint but points out that some imperfections in the original copies may be apparent.

Disclaimer
The publisher has made every effort to trace copyright holders and welcomes correspondence from those they have been unable to contact.

ISBN 13: 978-1-315-89827-8 (hbk)
ISBN 13: 978-1-351-07737-8 (ebk)

Visit the Taylor & Francis Web site at http://www.taylorandfrancis.com and the
CRC Press Web site at http://www.crcpress.com

Dedicated
to my Parents and Mentors
especially Profs. K. Hanumantha Rao and K. V. Jagannadha Rao

FOREWORD

Ever since the increased use of synthetic pesticides during the past 40 years, it has become apparent that these highly toxic materials did not only affect target organisms but often harmed nontarget organisms like fish and other aquatic life. Some of these adverse effects were so subtle that they were not recognized until analytical and biological testing methods had been refined to discover heretofore unrecognized deleterious effects on fish. More recently, the attention of fish biologists has been turned towards nonpesticidal industrial pollutants, like polychlorinated biphenyls (PCBs) and dioxins, which in the past were often mistaken for DDT and related pesticides.

Although there are a number of books and reviews on the subject of pesticide toxicity on fish, the present treatise by Dr. A. S. Murty is the most comprehensive work on this subject. Dr. Murty, who has carried out basic research on the effect of pesticides on fish in his native land of India and Germany, has gathered the work of over 1800 papers published world-wide and critically placed them in the context of 12 chapters. This thorough coverage of the world literature attests to the fact that no part of the globe is too remote to obtain the current scientific literature on a particular technical subject, as for example the toxicity of pesticides to fish. This present work is a detailed treatment of the environmental fate of pesticides and their acute and chronic toxicological effect on fish. A shorter chapter on PCBs and related compounds enhances the coverage of the book. An interesting treatment on insecticide resistance in fish and the combined action of different pesticides, to the best of my knowledge, have not been discussed as thoroughly as in this book.

The fluent style and the comprehensive coverage of this book make it not only a pleasure to read but also a "must" on the shelf or desk of concerned scientists all over the world.

Gunter Zweig, Ph.D.
Washington, D.C.
March, 1985

PREFACE

The increasing awareness of the environmental problems during the last 2 decades kindled much interest on the environmental fate and behavior of the xenobiotic chemicals. This is evident from the fact that during the last 16 years or so, about 20 international journals have been launched to focus attention on environmental problems. Environmental disasters like the collapse of the coho salmon fishery in Lake Michigan and the mass mortality of fish elsewhere clearly established the toxicity of pesticides to fish and other aquatic organisms. Further, the realization that the aquatic environment is the ultimate sink for all pollutants stimulated studies on the toxicity of man-made chemicals to the aquatic organisms, especially pesticide toxicity to fish. Much of the information on the toxicity of pesticides to fish, however, lay scattered in diverse journals or in generalized summaries on the effect of pollutants on fish. The present work is the first major attempt at reviewing comprehensively all the available information on this fascinating subject.

I have attempted to trace the growth of this subject from the earliest toxicity tests that I could record, dating back to 1944, to the present day. In about 4 decades, the toxicity testing itself has undergone a conceptual change. In the early days, simple toxic effects with death as the end point were studied. At present the focus has shifted to hazard assessment and prediction to avert the environmental disasters caused by DDT-type compounds.

In dealing with the toxicity of pesticides to fish, one may feel that perhaps a discussion on polychlorinated biphenyls (PCBs) and other industrial organic chemicals is out of place. Although such chemicals have rarely been, if ever, used in combination with pesticides, their environmental behavior and uptake by fish are akin to those of the organochlorine pesticides. Hence, the similarity between the two groups of compounds is briefly pointed out and the methods of their separation are discussed.

This book is the result of the constant encouragement I received from Dr. Gunter Zweig of the U.S. Environmental Protection Agency, Washington, D.C. To him, I owe a deep debt of gratitude. I also thank him for readily agreeing to write the Foreword for this book.

The many discussions I had with the fellow scientists, during my tenure of work as a visiting scientist of the International Development Research Centre (IDRC), Canada, during 1980 to 1981 and as a Fellow of the Alexander Von Humboldt Foundation of West Germany during 1981 to 1983 helped me in planning and completing this work. Especially IDRC had funded my visit to a number of laboratories. I thank both these agencies for affording me an opportunity to meet my fellow scientists. I also thank my parent University for granting me leave to go abroad and work.

Mr. J. R. W. Miles and Dr. S. Sethunathan read critically Chapter 1, Volume I. Drs. Richard Addison and J. L. Hamelink commented upon Chapter 3, Volume I, and Dr. D. Desaiah, upon the part dealing with ATPase-inhibition studies in Chapter 6, Volume II. I am obliged to them for their comments and suggestions. I would like to thank Drs. D. V. Subba Rao, S. Pitchumani, Bhuwanesh Agrawal, and U. Schiecke, Ms. Seetha Murty, and the librarian of the Columbia National Fisheries Research Laboratory, Columbia, Missouri, for their help with the necessary reprints. I also wish to acknowledge the help extended by Messers K. Kondaiah, K. Krishna Prasad, M. V. Krishnaiah, and K. R. S. Sambasiva Rao, in the preparation, finalization, and typing of the manuscript. The excellent editorial help of the staff of CRC Press is also acknowledged.

The sources of the figures and tables have been mentioned at the appropriate place. Besides, I would like to acknowledge Drs. R. D. Wauchope, J. L. Hamelink, G. D. Veith, H. C. Alexander, W. B. Neely, and C. J. Schmitt for the original photographs of their graphs.

I am also indebted to my colleagues Dr. Y. Ranga Reddy and Prof. S. Krishna Sarma, who cheerfully read through the entire manuscript and commented upon style and syntax. The faults, if any, are mine.

This work would not have been possible but for the involvement of my wife, Ramani, who supervised the different stages of the preparation and typing of the draft, and patiently corrected the proofs. Lastly, it is indeed a pleasure to acknowledge the help of our children, Usha, Sudha, and Aditya in classifying the literature and organizing the references.

A. S. Murty

THE AUTHOR

Dr. Ayyagari S. Murty, Reader in Zoology at Nagarjuna University, Nagarjunanagar, India is an environmental toxicologist interested in the toxicity of xenobiotic chemicals to nontarget organisms. He received his M.Sc. and Ph.D. in Zoology from Andhra University, India. As a teacher for postgraduate students for more than 20 years, Dr. Murty has taught diverse subjects ranging from systematic zoology to environmental biology.

Dr. Murty is presently engaged in research on environmental hazard evaluation of chemicals and also on the toxicity of pesticides and industrial organic chemicals to nontarget organisms, especially fish. During 1980 and 1981, he was one of the 18 scientists selected from the third-world countries in a world-wide competition by the International Development Research Centre and worked as a visiting scientist at London (Ontario) research station of Agriculture Canada. From 1981 to 1983, he worked as a Fellow of the Humboldt Foundation of West Germany, mostly at the Institute for Water, Soil, and Air Health of the Federal Health Office, West Berlin.

Dr. Murty presented several papers at national and international conferences. He is frequently invited to lecture at workshops and seminars on environmental toxicology. He published over 40 research papers and his work has brought to light certain lacunae in aquatic toxicity testing protocols with hydrophobic compounds.

TOXICITY OF PESTICIDES TO FISH

A. S. Murty

Volume I

Volume II

TABLE OF CONTENTS

Volume II

Chapter 1

TOXICITY OF INDIVIDUAL COMPOUNDS, SAFE CONCENTRATIONS, AND TOXICITY TO DIFFERENT AGE GROUPS

I. INTRODUCTION

Since the early times of their use, it has been recognized that the synthetic, organic pesticides are extremely toxic to fish and other aquatic organisms. Hence, much work on toxicity testing (mostly in the form of identifying the concentration of a compound that would kill 50% of the test fish in 24 to 96 hr) ensued, and voluminous literature on the acute toxicity of pesticides to fish accrued. It would be an unwieldy task to review the available literature on the acute lethality of pesticides to fish. The reader is referred to the excellent reviews on this subject by Johnson,[1] Alabaster,[2] Pimentel,[3] Johnson,[4] and Holden.[5] The emphasis in this chapter is laid on the work published after 1970, as it was only then that a greater degree of standardization of the methods for conducting toxicity tests was achieved. In this context, reference should be made to the compilation of the data on all the toxicity tests conducted at the Columbia National Fisheries Research Laboratory (CNFRL), Missouri. Johnson and Finley,[6] in this excellent compilation, summed up the result of 1587 tests conducted over a period of 14 years at CNFRL, with 271 chemicals and 58 test species (28 species of fish and 30 species of invertebrates). This compilation is the single largest source of dependable information on the acute toxicity of many pesticides to fish and invertebrates, the tests having been conducted under nearly uniform conditions.

II. TOXICITY OF DIFFERENT PESTICIDES

A. Acute Toxicity
1. Organochlorines
In general, organochlorine (OC) compounds are more toxic than organophosphate (OP) compounds to fish, although a few of the latter are as toxic as some of the highly toxic OC compounds. Among the OC, endrin and other cyclodienes are highly toxic to fish. The 96-h LC 50 of technical endrin to various species of freshwater fish ranged from 0.7 to 2.1 $\mu g/\ell$ under static conditions,[7] and 0.4 to 0.63 $\mu g/\ell$ to the estuarine fish under continuous flow conditions.[8] The toxicity of endrin to the freshwater organisms was reviewed by Grant.[9]

Endosulfan, another cyclodiene compound, is also highly toxic to fish. Its toxicity to several species of freshwater fish is in the range of 0.2 to 8.1 $\mu g/\ell$,[10-14] and to the No. American saltwater fish it is in the range of 0.3 to 2.9 $\mu g/\ell$.[10] In nature, fish were able to tolerate a short-term exposure to four times the concentration of endosulfan that killed all the fish in 24 hr.[15] Heptachlor is another cyclodiene with high toxicity to fish, its 96-h LC 50 for several species of estuarine fish being 1 to 4 $\mu g/\ell$.[16] The 96-h LC 50 of chlordane was 6.4 and 24.5 $\mu g/\ell$ to pinfish and sheepshead minnows.[17] Of the photoisomerization products of cyclodienes, photoaldrin and photodieldrin are more toxic than their normal isomers (whereas photoisodrin is less toxic than isodrin[18]). The greater toxicity of the first two photoisomers is due to their in vivo conversion to more toxic ketones. Likewise, photoheptachlor is more toxic, whereas photochlordane is less toxic than their respective parent compounds.[19] Toxaphene, a chlorinated camphene, is also highly toxic to fish, the 96-h LC 50 to estuarine fish being < 1 $\mu g/\ell$.[20] Of the many components of technical toxaphene, heptachlorobornane was identified as the most toxic.[21] The acute toxicity of DDT to fish is less than that of many other OC compounds. Of the two isomers of DDT, the toxicity of *o,p'*-DDT is about one third that of *p,p'*-DDT.[2] Closely related methoxychlor and its analog — dianisylneopentane — are much less toxic than DDT, the 96-h LC 50 to *Gambusia* being 1, 3, and 0.32 mg/ℓ, respectively.[22]

The reported 96-h LC 50 of lindane and technical hexachlorocyclohexane (HCH) to several species of estuarine fish was reported to be 30 to 104 μg/ℓ; however, their toxicity to crustaceans was very high (< 10 μg/ℓ). The chronic toxicity of mirex was more than that of Kepone®, whereas the reverse was true for short-term toxicity.[23] The acute toxicity of hexachlorobenzene (HCB) to fish is negligible, but its chronic toxicity is high because of its high potential for bioaccumulation. In light of the reports that great quantities of this persistent compound enter the environment, not because of its use as a pesticide, but because of its release as a by-product of many industrial processes involving chlorination, the environmental persistence and long-term toxicity of HCB are causes for concern.[24]

2. Organophosphates, Carbamates, and Pyrethroids

OPs have negligible chronic toxicity, but some of them have moderate to high acute toxicity. The 96-h LC 50 of several OP compounds to various species of No. American freshwater fish, as recently reported, are[6]

1.   Ethyl parathion, < 1 to 3.3 mg/ℓ
2.   Ethion, 0.5 to 7.6 mg/ℓ
3.   Guthion® (ethyl), 1.1 to 20 μg/ℓ and Guthion® (methyl), 0.4 to 4300 μg/ℓ (but for most species less than 50 μg/ℓ)
4.   Acephate, 100 to more than 1000 mg/ℓ
5.   Chlorpyrifos, 2.4 to 280 μg/ℓ
6.   Fenitrothion, 2 to 12 mg/ℓ
7.   Phorate, 6 to 280 μg/ℓ

The compounds that have a P = S linkage have very low, direct toxicity; but they prove to be highly toxic upon being converted to their P = O analogs. The acute toxicity of a few OP compounds was less for air-breathing fish, owing to the reduced intake of the toxicant via gills.[25] Delayed mortality with no deaths at the end of 96 hr but 100% mortality at the end of 9 days, was observed with coumaphos.[26]

Carbamates are moderately toxic to fish, but highly toxic to invertebrates (the 48- or 96-h EC 50 for many invertebrates is less than 10 μg/ℓ). Carbaryl, the more commonly used carbamate, has a 96-h LC 50 of 2 to 39 mg/ℓ and carbofuran 150 to 870 μg/ℓ, for many No. American freshwater fish.[6]

Pyrethroids have very high insecticidal activity and have found wide use in the last few years. Though they are not persistent in the environment, their acute toxicity to fish is high.[27] The 96-h LC 50s of a synthetic pyrethroid — SBP-1382 — is in the range of 1 to 5 μg/ℓ, for several salmonids.[28] The cyano-substitution of the phenoxybenzyl alcohol moiety (as in cypermethrin and fenvalerate) enhances the toxicity of pyrethroids to fish. Similarly, pyrethroids with (1R)-cis-isomers of the acid moiety are more lethal than the corresponding cis-, trans-racemates.[29] Fenvalerate and permethrin are more toxic than many OC compounds, the 96-h LC 50 being 0.5 to 12.0 μg/ℓ.[30] [32] Cold water fish were reportedly more susceptible than warm water fish to natural as well as synthetic pyrethroids.[33]

3. Herbicides

Few of the herbicides have any chronic toxicity. The acute toxicity of the herbicides is also low. The major problem arising from the application of herbicides for the control of aquatic weeds is not direct but results from the depletion of dissolved oxygen owing to the decomposition of the dead and decaying weeds. Another problem with herbicides is the very high quantities that have to be initially used for effective weed control. The toxicity of many herbicides to plants as against the aquatic animals is less by several orders of magnitude.[34,35]

Out of 19 compounds tested for their toxicity to 1-year old fingerling channel catfish, only 4 compounds were of some interest. The 96-h LC 50 values of these four were (all values mg/kg) (1) DNBB, 0.12; (2) bensulide, 0.38; (3) trifluralin, 0.42; (4) propanil, 3.8; at a pH of 8.2. Other herbicides were not toxic even at 10 mg/kg.[36]

The toxicity of 2,4-D (2,4-dichlorophenoxyacetic acid) depends on the type of formulation used and the lethal concentration ranged from 5 to 1500 μg/ℓ.[37] Among the other herbicides, cresols are more toxic than phenols.[38] The toxicity of pentachlorophenol (PCP) has been reviewed.[39] The 96-h LC 50 of sodium pentachlorophenate to an air-breathing fish was 0.39 mg/ℓ.[40] In nature the phenolic form is more persistent than the phenate salt.[41] The 96-h LC 50 of PCP to fathead minnows was 0.22 mg/ℓ.[42] The toxicity (96- and 192-h LC 50) of a number of substituted phenols to the fathead minnow was reported by Phipps et al.[42] MSMA (monosodium methanearsonate), an arsenical compound, has very low toxicity to black bass and catfish.[43] The 96-h LC 50 of propamocarb to carp, bluegill, and rainbow trout was, respectively, 234, 415, and 616 mg/kg.[44] On the other hand, captan, a fungicide, is highly toxic, the 96-h TL 50 being 34 to 72 μg/ℓ for several No. American freshwater fish.[45] Glyphosate or Roundup®,[46] Asulam,[47] Bolero® 8EC[48] (a carbamate, pre-emergence herbicide) were not at all toxic, while dinitramine was reported to be moderately so.[49]

Hydrogen ion concentration (pH) influences the toxicity of herbicides, as most of them are ionizable. The unionized state of the compound has greater toxicity than the dissociated components and the degree of ionization is dependent on the pH. The influence of pH on the toxicity of herbicides is considered in the next chapter. Although the toxicity of herbicides to fish and invertebrates is much less than that of many insecticides, relatively high concentrations of the former are needed for effective weed control. Furthermore, many of these herbicides are applied directly to the aquatic ecosystem at field rates that result in water concentration in excess of the median toxic concentration for *Daphnia* and other fish food organisms.[50] Hence it is through their effect on fish food organisms that herbicides may affect fish.

4. Predicting the Acute Toxicity

Zitko et al.[32] found a good correlation between octanol/water partition coefficient (log K_{ow} or log$_P$) and the lethality of a number of pyrethrins, which is described by the following equation:

$$Y = 0.422 \ \Sigma \ \log_P - 0.7, \text{ with} = 0.924$$

No such relationship could be established in the case of cyano-substituted pyrethroids, which are more toxic than what was predicted by the above relationship.[51]

Kanazawa[52] attempted to find a correlation between bioconcentration (BCF) and toxicity of several compounds. He found a satisfactory relationship between log (toxicity) to carp and log (BCF) of topmouth gudgeon after excluding the LC 50 values that exceeded the water solubility of the respective compounds. The relationship is described by the following equation:

$$\log Y_1 = 5 - (0.764 \log X)$$

where Y_1 is the acute toxicity to carp (48-h LC 50 in μg/ℓ) and X is the BCF by topmouth gudgeon. But no significant relationship between log toxicity to topmouth gudgeon and log (BCF) by the same fish was observed, leaving such suggested relationships open to criticism.

B. Chronic Toxicity and Safe Concentrations

The ultimate aim of toxicity testing is to predict the acceptable levels of toxicant in the environment, which will be harmless to the biota. The various types of chronic tests, viz., 30-day tests, multigeneration and whole life cycle tests, partial life cycle and embryo-larval tests and the like, reviewed in Chapter 4, Volume 1, aim at identifying a concentration of the toxicant that may be considered harmless or "safe" and does not produce any adverse sublethal effects. As has been emphasized by Sprague,[53] the term "safe level" is not an entirely satisfactory one, but in the absence of a better term, it has gained wide acceptance. In the context of a supposed safe level, an application factor (AF) is used, which is obtained by multiplying the LC 50 by an arbitrary factor such as 0.1, to obtain a concentration that is supposed to cause no adverse sublethal effects. For instance, Hart et al.[54] proposed the following formula

$$C = \frac{48\text{-h LC} \times 0.3}{S^2}$$

where C is the presumably harmless concentration and

$$S = \frac{24\text{-h LC 50}}{48\text{-h LC 50}}$$

Another approach to arrive at safe level is to use the one tenth of the 48-h LC 50 value.[55] The European Fisheries Advisory Commission (EFAC)[56] suggested that a concentration less than four orders of magnitude lower than the 96-h LC 50 is an acceptable or safe concentration. Lloyd[57] opined that this approach considerably over-estimates the likely hazard of a chemical. The International Joint Commission adopted a water Quality Objective for nonpersistent pesticides, which is 5% of the 96-h LC 50 value to the most sensitive fish species.[58] Sprague[53] suggested the use of EC_5 or EC_1 (calculated by the extrapolation of LC 50 or EC 50), as against a no-effect concentration, to make the safe-concentration independent of the arbitrariness, in choosing test concentrations. Stephan and Mount[59] reviewed the various attempts to calculate the AF.

Mount and Stephan[60] underscored the fact that extrapolation of laboratory data to the field is not always meaningful and hence it is difficult to decide upon an acceptable concentration based on the laboratory experiments that may be considered "safe" in the field, too. The problem is further complicated by the fact that many aquatic toxicologists have not agreed upon the type of effect that is to be measured to assess damage. Mount and Stephan proposed to use Laboratory Fish Production Index (LFPI), which incorporates data generated over at least one generation of the test species, regarding effects on survival, growth, reproduction, spawning behavior, and viability of eggs and growth of fry, to calculate the Maximum Acceptable Toxicant Concentration (MATC) that would not affect fish production in nature. It may be recalled that MATC is a range of concentration. The lower end of MATC is the concentration that does not produce any adverse effect. The upper end of MATC is the lowest concentration that would cause an adverse effect on some life stage or the other. To ensure that all life stages of species are exposed, the test is begun with eggs or 20-day old or younger fry and continued until the offspring of these fish are at least 30 days old. Although such MATC values may have certain drawbacks, they give more meaningful data than the simple extrapolation of a 96-h LC 50 value. Further, MATC calculated for species amenable to laboratory testing can be used for establishing safe concentrations for other species that cannot be tested in the laboratory. Mount and Stephan proposed that the AF be calculated by dividing the MATC by the 48- or 96-h LC 50 value. By using the AF of a compound for one species and the LC 50 of the same to a second species, the MATC for the second species may be calculated. Since a large data base on the acute toxicity of all xenobiotic compounds exists, a carefully calculated MATC for one species can be used to furnish much more meaningful information. For instance, despite the fact that the long-term toxicity of malathion to bluegills is more than 50 times that to the fathead minnow, the safe AF calculated for both species (MATC ÷ 96-h LC 50) were similar and very close to each other.[61] It should be underlined, however, that the accuracy of the MATC can be no better than the determination of the closeness of the concentrations that cause and that do not cause any adverse effect.

The use of AF for calculating safe levels for other species was explained and discussed in detail by Eaton.[61] Since Mount and Stephan[60] proposed MATC as a meaningful measure to calculate a safe AF, many studies have been conducted to report the MATC of several compounds to a number of species.[45,62]

Eaton[63] recommended that the lethal threshold concentration (LTC, the point in the acute toxicity test at which the mortality is 10% or less of the original number of fish in any concentration during the preceding 24 hr, is more appropriate than the LC 50 values for calculating the MATC. Johnson and Julin[64] considered that the 96-h LC 50 values would be as acceptable as LTC since they agree with each other, and the former data are less time-consuming, less expensive to generate, and are widely accepted. They recommended the continued use of 96-h LC 50 for calculating the MATC values, when experimentally determined MATCs are not available.

Kenaga[62] pointed out that the AFs of some compounds for fish may vary by as much as 10,000 to 100,000 times. Further, the MATC values of several chemicals to the same species vary by more than 1000-fold; for the same compound among different species, the variation is more than 100-fold. He emphasized that the major weakness in the calculation of AF is its dependence on the LC 50 value. While the latter is based on lethality, the MATC is based on the use of unspecified nonlethal properties, and, hence, strictly speaking, the LC 50 and MATC are not equivalent and comparable values. Hence, Kenaga developed interspecific regression equations for predicting the AFs for one species, based on those of another.

The AFs have to be used with caution. Although an AF of 0.01 has been recommended for Aroclor® 1254, and the calculated MATC was ≥ 1 $\mu g/\ell$, Mauck et al.[65] pointed out that at a concentration below the MATC, sublethal backbone collagen changes were evident. This study indicated that the AF of 0.01 for Aroclor® 1254 will not be correct.

McKim[66] compared 56 fish life cycle tests involving 34 chemicals and 4 species of fish and showed that the embryo-larval tests offer the same type of information, with comparable dependability as the long-term chronic tests, and are at the same time cost-efficient and time-saving (Table 1).

While a chronic whole life cycle or partial life cycle test requires about a year, a whole life cycle test with *Daphnia* takes only 3 weeks to complete. Maki[67] compared the concentrations of six surfactants that did not cause any adverse effects to *Daphnia* in a whole life cycle test, with MATC to the fathead minnow. He found a very high correlation between the two (r = 0.98), and recommended that a 21-day *Daphnia* chronic test may be used instead of whole life cycle fish tests.

Table 1

MAXIMUM ACCEPTABLE TOXICANT CONCENTRATIONS
(MATC) FROM PARTIAL AND COMPLETE LIFE CYCLE
TOXICITY TESTS WITH FISH AS COMPARED WITH MATCs
DERIVED FROM EMBRYO, LARVAL, AND EARLY JUVENILE
TOXICITY TESTS

Toxicant	Fish species	Partial/complete life cycle MATCs ($\mu g/l$)	Embryo-larval/ juvenile MATCs ($\mu g/l$)
Pesticides			
Acrolein	Fathead minnow	11—42	11—42
Atrazine	Brook trout	60—120	120—240
Trifluralin	Fathead minnow	2.0—5.1	5.1—8.2
Endosulfan	Fathead minnow	0.20—0.40	0.20—0.40
Endrin	Flagfish	0.22—0.30	0.22—0.30
Heptachlor	Fathead minnow	0.86—1.8	0.86—1.8
Diazinon	Flagfish	54—88	54—88
	Fathead minnow	6.8—14	6.8—14
Guthion	Fathead minnow	0.33—0.51	0.70—1.8
Malathion	Flagfish	8.6—11	8.6—11
Polychlorinated biphenyls			
Aroclor® 1242	Fathead minnow	5.4—15	5.4—15
Aroclor® 1248	Fathead minnow	1.1—3.0	1.1—4.4
Aroclor® 1254	Fathead minnow	1.8—4.6	1.8—4.6
Aroclor® 1260	Fathead minnow	2.1—4.0	2.1—4.0
Metals			
Cadmium	Flagfish	4.1—8.1	8.1—16
	Fathead minnow	37—57	37—57
Chromium	Fathead minnow	1,000—3,950	1,000—3,950
Copper	Brook trout	9.5—17	9.5—17
	Fathead minnow	11—18	11—18
Lead	Brook trout	58—119	58—119
	Flagfish	31—62	62—125
Nickel	Fathead minnow	380—730	380—730
Zinc	Flagfish	26—51	51—85
	Fathead minnow	30—180	30—180

Data from McKim, J. M., *J. Fish. Res. Board Can.*, 34, 1148, 1977; as condensed by Mayer, F. L., Jr., Mehrle, P. M., Jr., and Schoettger, R. A., Trends in aquatic toxicology in the United States: a perspective, EPA-600/9-80-034, Proc. 3rd U.S.-U.S.S.R. Symp. Effects of Pollutants upon Aquatic Ecosystems: theoretical aspects of aquatic toxicology, U.S. Environmental Protection Agency, Duluth, Minn., 1980, 241. With permission.

III. RELATIVE TOXICITY OF PESTICIDES

While testing the toxicity of a number of compounds to one or more species, or the toxicity of the same compound to a number of species at the same time, the question often arises as to which of these compounds is most toxic or which of these species is the most sensitive or the most tolerant.

Katz[26] tested the toxicity of 13 compounds to 4 species of fish. Out of nine OC, three OP, and one carbamate compounds tested, endrin was the most toxic to fish. Coumaphos (Co-Ral®), an OP compound, was the least toxic of the 13 compounds. Coho salmon and rainbow trout, in general, were more tolerant to these insecticides than chinook salmon. Pickering et al.[68] evaluated the toxicity of 13 OP compounds to 4 species of fish. Bluegills were generally a very sensitive species, followed by guppies, fathead minnows, and goldfish. Guthion® and EPN were the most toxic compounds.

In an evaluation of the toxicity of ten OC compounds to 4 species of fish, Henderson et al.[7] reported that endrin was the most toxic and HCH was the least toxic. For nine compounds, the 96-h LC 50 was below 100 $\mu g/l$. Macek and McAllister[69] chose 12 species of fish belonging to 4 families, viz., Ictaluridae, Cyprinidae, Centrachidae, and Salmonidae to test the relative toxicity of nine insecticides, viz., four OPs, three OCs, and two carbamates. Members of the same family showed somewhat similar susceptibilities to the different compounds; salmonids were the most susceptible and ictalurids and cyprinids, the least susceptible. Brown trout was the most sensitive species to OCs, and coho salmon, the most sensitive to carbamates; goldfish was the least sensitive to any group of compounds. The difference in the sensitivity of the most sensitive and the least sensitive species to phosphorothionates was not much, whereas in the case of the phosphorodithioates, it was considerable. Reviewing the published literature at that time, Macek and McAllister examined the chemical structure of the pesticide molecule, rate of formation of P = O analogs of compounds having a P = S linkage, and the extent of tolerance of the inhibition of acetylcholinesterase (AChE) by the different species as the possible causes for this difference in sensitivity. They concluded that toxicity tests with relatively insensitive species like goldfish should be discouraged while attempting the evaluation of the environmental hazards of a compound.

Post and Schroeder[70] studied the toxicity of four insecticides (DDT, endrin, malathion, and carbaryl) to four species of No. American freshwater fish. Eisler[71] employed a rank correlation method to determine the order to toxicity of 12 insecticides (7 OCs and 5 OPs) to 7 species of fish. Confirming the earlier reports, he concluded that endrin was the most toxic, and methyl parathion, the least toxic compound; the OP compound dioxathion was more toxic than many OC compounds.

In a similar study, the overall degree of toxicity of several classes of compounds to bluegills was summarized as follows: toxicity of chlorinated phenols < chlorinated benzenes < chlorinated ethylenes < chlorinated ethanes.[72]

The relative sensitivity of several species of fish to DDT was summarized by an EPA (U.S. Environmental Protection Agency) report,[73] where the LC 50 values varied from 0.6 to yellow perch to 180 $\mu g/l$ to the goldfish (this latter value may be viewed against a known water solubility of 1.2 $\mu g/l$ of DDT).[73]

To evaluate the relative potency of a large number of toxicants tested against the same or different organisms, Marking[74] proposed a formula, to evaluate effectively and rapidly the toxicity data. With the help of this formula, the evaluation of the relative potency is also possible. The potency rating (Pr) is calculated by

$$Pr = \frac{\sum_{j=1}^{5} \sum_{i=1}^{3} \frac{M(C_i, T_j)}{C_i \log T_j}}{\log N}$$

where M = mortality among the total number tested; C = concentrations; T = time of exposure in hours, T > 1; N = number of observations, N > 1; j = finite variables of T: $j_1 = 3$, $j_2 = 6$, $j_3 = 24$, $j_4 = 48$, $j_5 = 96$ (but any other series of time in hours may be taken;[93] i = finite variables of C: $i_1 = 0.1$, $i_2 = 1.0$, $i_3 = 10$.

The (Pr) helps separate toxic chemicals from nontoxic ones. It also indicates the general toxicity and differential toxicity among the various species tested. Using this formula, the results of tests of different durations can also be compared with one another. This (Pr) is more applicable to chemicals which have similar dose-effect curves. Marking and Willford[75] evaluated and reported the relative toxicity of 29 nitrosalicylanilides using the above method.

IV. TOXICITY OF PESTICIDES TO DIFFERENT AGE GROUPS

It is well known that different life stages of fish differ in the degree of their sensitivity to the toxicants. More commonly, the eggs are the least sensitive, along with larvae with yolk sacs. With the regression of the yolk sac, the toxicity increases manyfold to the young larva or the juvenile, but with increase in age, the toxicity decreases. The lesser sensitivity of the eggs is attributable to the relative impermeability of the egg membranes to xenobiotic compounds; the greater tolerance by the larger size/age groups is attributable to their larger lipid pool and also a decrease in the gill surface area in relation to the body size.

Iyatomi et al.[76] noted that the 24-h LC 50 value to egg, 1-, 4-, and 6-day-old larva of the carp was, respectively, 19.9, 8.5, 4.2, and 0.061 mg/kg. Similarly, the 24-h LC 50 value to snake-head fish eggs, 1-, 4-, and 12-day-old larva was 100, 62, 5.6, and 0.065 mg/kg, respectively. Parrish et al.[17] reported that although the survival of eggs and embryos of sheepshead minnows was not affected by exposure to chlordane, fry died in significant numbers in the 17 and 36 µg/ℓ concentrations. There was no difference in the development and hatching success between normal rainbow trout eggs and those that were exposed for 30 to 120 min to various concentrations of endosulfan (10 to 50,000 µg/ℓ).[11] In experiments with TFM (3-trifluoromethyl-4-nitrophenol), green eggs of coho salmon were the most sensitive, followed by fry, sacfry, and eyed eggs, the 96-h LC 50 being 0.64 mg/ℓ for green eggs and 3.5 mg/ℓ for the eyed eggs.[77] In another study, eggs of chinook salmon, brook trout, and lake trout were 15 times less sensitive to the toxicity of TFM than fingerlings.[78] Early life stages of rainbow trout and channel catfish were the least sensitive to glyphosate, a herbicide; the egg stage was the least sensitive and toxicity increased to the sacfry and early swim-up stages, but toxicity decreased as the fish grew larger.[79]

Decreasing toxicity with increase in age/size was observed in the case of TFM with rainbow trout and coho salmon,[80] and also in the case of Kepone® with fathead minnows.[81] Likewise, greater sensitivity of fry than the embryos of sheepshead minnows to the toxic action of toxaphene[20] and Aroclor® 1254[82] was reported. In a study on the acute toxicity of fenitrothion to rainbow trout, embryos were the least sensitive, sacfry stage was intermediate, and fingerlings and adults, in that order, were the most sensitive.[83] The influence of the size of the test fish on the toxicity of DDT given through diet was also reported by Buhler et al.[84] Smaller and younger fish were more susceptible to DDT than larger and older fish. Buhler and Shanks,[85] in another study, noted the decreasing toxicity of DDT to larger and older coho salmon, and explained it as the result of the decrease in the diet consumption per unit body weight by larger fish, in comparison with that of the smaller ones. Besides, the greater toxicity of DDT to smaller fish is also partly attributable to the lesser lipid content of smaller fish, which fails to provide adequate storage for the toxicant taken up.

Greater sensitivity of young striped mullet than older juveniles to mirex and methoxychlor,[86] was reported. Adult striped mullets had larger amounts of the residues, owing to the presence of a relatively higher percentage of body fat in the adults when compared with the juveniles. That it is the lipid content that confers a greater degree of tolerance on the larger fish to the DDT-type compounds is confirmed by the toxic concentrations of DDT, endrin, and malathion, to brook trout, cutthroat trout, and coho salmon of 1-g size and 1- to 2-g size, reported by Post and Schroeder.[70] With DDT and endrin, there was reduction in the toxicity with increase in size, whereas no such difference in the toxicity of malathion to smaller and larger fish was found. The 96-h LC 50 of parathion to rainbow trout weighing 1 g was 3.17 µg/ℓ, whereas for a larger size group (200 g), it was 314 µg/ℓ.[87] A comparable finding with Delnav®, with no difference in its toxicity to 1- or 38-day old fathead minnows, and with tetraethyldiphosphate (TEPP), with 3-day old fry and adult fathead minnows, was reported by Pickering et al.[68] But in the same study, greater toxicity of Delnav®, parathion, and malathion to the smaller bluegills than larger ones was also reported. Gupta et al.[88] found that in general the toxicity of phenol, 2,4-dinitrophenol and PCP decreased with increase in the size of the fish; however, the largest size fish were more sensitive than those belonging to the next smaller size group. The greater sensitivity of smaller guppies than larger guppies to TCDD (tetrachloro dibenzo dioxin) poisoning was perhaps the result of the larger gill surface area to body mass ratio in the smaller fish in comparison with the larger fish, permitting greater uptake in the former.[89]

Adelman et al.,[90] while discussing the utility of PCP as a reference toxicant, observed no difference in the LC 50 to smaller and larger size groups of fathead minnows, except in the early stages of the test. They concluded that young fish may be more conveniently and profitably tested than the larger ones.

Influence of sex on the toxicity of a compound has been reported in a few studies. Fathead minnows that died following exposure to DDT, either in food or in water or both, were predominantly males. The residues in the dead males, however, were only 55% of those still surviving. Lipid percentage of the dead males was significantly lower than that of the surviving males.[91] Significant difference in the toxicity of phenols to male and female guppies was observed.[92] Greater susceptibility of male rainbow trout to endosulfan was noted by Schoettger,[11] but the author also noted that the males were smaller specimens and hence their higher mortality cannot be attributed to sex alone.

V. CONCLUSIONS

Among the OC compounds, cyclodienes have high acute toxicity, whereas DDT has a greater chronic toxicity. Some DDT analogs have low acute and chronic toxicity to fish. The chronic toxicity of OP and carbamate compounds is low, but some of the former are as toxic as some of the most toxic OC. The toxicity of herbicides is low, but their effect seems indirect, i.e., depletion of oxygen by decaying vegetation following the application of herbicides for the control of aquatic weeds.

Among the different life stages of fish, the eggs and larvae with yolk sac are the least susceptible to pesticides. The feeding larvae and juveniles are the most sensitive. Toxicity decreases as the fish grows.

The greater toxicity of pesticides to the younger stages of fish appears to be related to (1) a higher rate of metabolism and hence a higher food intake per unit body weight in the young fish, (2) greater intake of the toxicant via gills, because of the larger gill surface area relative to body mass in the smaller fish, and (3) smaller lipid pool of the body that can store only smaller quantities of the toxicant.

REFERENCES

1. Johnson, D. W., Pesticides and fishes — a review of selected literature, *Trans. Am. Fish. Soc.*, 97, 398, 1968.
2. Alabaster, J. S., Survival of fish in 164 herbicides, insecticides, fungicides, wetting agents and miscellaneous substances, *Int. Pest Control*, 2, 29, 1969.
3. Pimentel, D., Ecological Effects of Pesticides on Non-target Species, Executive Office of the President's Office of Science and Technology, U.S. Government Printing Office, Washington, D.C., 1971.
4. Johnson, D. W., Pesticide residues in fish, in *Environmental Pollution by Pesticides*, Edwards, C. A., Ed., Plenum Press, New York, 1973, 5.

5. Holden, A. V., Effects of pesticides on fish, in *Environmental Pollution by Pesticides*, Edwards, C. A., Ed., Plenum Press, New York, 1973, 6.

6. Johnson, W. W. and Finley, M. T., *Handbook of Acute Toxicity of Chemicals to Fish and Aquatic Invertebrates*, Resource Publ. 137, Fish and Wildlife Service, U.S. Department of the Interior, Washington, D.C., 1980.

7. Henderson, C., Pickering, Q. H., and Tarzwell, C. M., Relative toxicity of ten chlorinated hydrocarbon insecticides to four species of fish, *Trans. Am. Fish. Soc.*, 88, 23, 1959.

8. Schimmel, S. C., Parrish, P. R., Hansen, D. J., Patrick, J. M., and Forester, J., Endrin: effects on several estuarine organisms, Proc. 28th Annu. Conf. Southeastern Assoc. Game Fish Comm., White Sulphur Springs, W. Va., 1975, 187.

9. Grant, B. F., Endrin toxicity and distribution in freshwater: a review, *Bull. Environ. Contam. Toxicol.*, 15, 283, 1976.

10. Ambient Water Quality Criteria for Endosulfan, EPA 440/5-80-046, Office of Water Regulations and Standards Criteria and Standards Division, Environmental Protection Agency, Washington, D.C., 1980.

11. Schoettger, R. A., Toxicology of Thiodan in several fish and aquatic invertebrates, in Investigations in Fish Control, Bureau of Sport Fisheries and Wildlife, Fish and Wildlife Service, U.S. Department of the Interior, Washington, D.C., 1970, 1.

12. Rao, D. M. R. and Murty, A. S., Toxicity and metabolism of endosulfan in three freshwater catfishes, *Environ. Pollut.*, 27, 223, 1982.

13. Priyamvada Devi, A., Rao, D. M. R., Tilak, K. S., and Murty, A. S., Relative toxicity of the technical grade material, isomers, and formulations of endosulfan to the fish *Channa punctata*, *Bull. Environ. Contam. Toxicol.*, 27, 239, 1981.

14. Ananda Swarup, P., Mohana Rao, D., and Murty, A. S., Toxicity of endosulfan to the freshwater fish *Cirrhinus mrigala*, *Bull. Environ. Contam. Toxicol.*, 27, 850, 1981.

15. Gorbach, S., Haarring, R., Knauf, W., and Werner, H. J., Residue analyses and biotests in rice fields of East Java, treated with Thiodan, *Bull. Environ. Contam. Toxicol.*, 6, 193, 1971.

16. Schimmel, S. C., Patrick, J. M. Jr., and Forester, J., Heptachlor: toxicity to and uptake by several estuarine organisms, *J. Toxicol. Environ. Health*, 1, 955, 1976.

17. Parrish, P. R., Schimmel, S. C., Hansen, D. J., Patrick, J. M. Jr., and Forester, J., Chlordane: effects on several estuarine organisms, *J. Toxicol. Environ. Health*, 1, 485, 1976.

18. Georgacakis, E., Chandran, S. R., and Khan, M. A. Q., Toxicity-metabolism relationship of the photoisomers of cyclodiene insecticides in freshwater animals, *Bull. Environ. Contam. Toxicol.*, 6, 535, 1971.

19. Khan, H. M. and Khan, M. A. Q., Biological magnification of photodieldrin by food chain organisms, *Arch. Environ. Contam. Toxicol.*, 2, 289, 1974.

20. Schimmel, S. C., Patrick, J. M., Jr., and Forester, J., Uptake and toxicity of toxaphene in several estuarine organisms, *Arch. Environ. Contam. Toxicol.*, 5, 353, 1977.

21. Saleh, M. A., Turner, W. V., and Casida, J. E., Polychlorobornane components of toxaphene: structure-toxicity relations and metabolic reductive dechlorination, *Science*, 198, 1256, 1977.

22. Coats, J. R., Metcalf, L., and Kapoor, I. P., Metabolism of the methoxychlor isostere, dianisylneopentane, in mouse, insects, and a model ecosystem, *Pestic. Biochem. Physiol.*, 4, 201, 1974.

23. Buckler, D. R., Witt, A., Jr., Mayer, F. L., and Huckins, J. N., Acute and chronic effects of Kepone and mirex on the fathead minnow, *Trans. Am. Fish. Soc.*, 110, 270, 1981.

24. Isensee, A. R., Holden, E. R., Woolson, E. A., and Jones, G. E., Soil persistence and aquatic bioaccumulation potential of hexachlorobenzene (HCB), *J. Agric. Food Chem.*, 24, 1210, 1976.

25. Jacob, S. S., Balakrishnan Nair, N., and Balasubramanian, N. K., Toxicity of certain mosquito larvicides to the larvivorous fishes *Aplocheilus lineatus* (Cuv. and Val.) and *Macropodus supanus* (Cuv. and Val.), *Environ. Pollut.*, 28, 7, 1982.

26. Katz, M., Acute toxicity of some organic insecticides to three species of salmonids and to the threespine stickleback, *Trans. Am. Fish. Soc.*, 90, 264, 1961.

27. Coats, J. R. and O'Donnell-Jeffery, N. L., Toxicity of four synthetic pyrethroid insecticides to rainbow trout, *Bull. Environ. Contam. Toxicol.*, 23, 250, 1979.

28. Marking, L. L., Toxicity of the synthetic pyrethroid SBP-1382 to fish, *Prog. Fish-Culturist*, 36, 144, 1974.

29. McLeese, D. W., Metcalfe, C. D., and Zitko, V., Lethality of permethrin, cypermethrin and fenvalerate to salmon, lobster and shrimp, *Bull. Environ. Contam. Toxicol.*, 25, 950, 1980.

30. Linden, E., Bengtson, B.-E., Svanberg, O., and Sundstrom, G., The acute toxicity of 78 chemicals and pesticide formulations against two brackish water organisms, the bleak (*Alburnus alburnus*) and the harpacticoid (*Nitocra spinipes*), *Chemosphere*, 8, 843, 1979.

31. Jolly, A. L., Jr., Avault, J. W., Jr., Koonge, K. L., and Graves, J. B., Acute toxicity of permethrin to several aquatic animals, *Trans. Am. Fish. Soc.*, 107, 825, 1978.

32. Zitko, V., Carson, W. G., and Metcalfe, C. D., Toxicity of pyrethroids to juvenile Atlantic salmon, *Bull. Environ. Contam. Toxicol.*, 18, 35, 1977.
33. Mauck, W. L. and Olson, L. E., Toxicity of natural pyrethrins and five pyrethroids to fish, *Arch. Environ. Contam. Toxicol.*, 4, 18, 1976.
34. Kenaga, E. E. and Moolenaar, R. J., Fish and *Daphnia* toxicity as surrogates for aquatic vascular plants and algae, *Environ. Sci. Technol.*, 13, 1479, 1979.
35. Frank, P. A., Herbicidal residues in aquatic environments, in *Fate of Organic Pesticides in the Aquatic Environment (Advances in Chemistry Series 111)*, Faust, S. D., Ed., American Chemical Society, Washington, D.C., 1972, 135.
36. McCorkle, F. M., Chambers, J. E., and Yarbrough, J. D., Acute toxicities of selected herbicides to fingerling channel catfish, *Ictalurus punctatus, Bull. Environ. Contam. Toxicol.*, 18, 267, 1977.
37. Sears, H. S. and Meehan, W. R., Short-term effects of 2,4-D on aquatic organisms in the Nakwasina River watershed, Southeastern Alaska, *J. Pestic Monit.*, 5, 213, 1971.
38. Hattula, M. L., Reunanen, H., and Arstila, A. U., The toxicity of MCPA to fish. Light and electron microscopy and the chemical analysis of the tissue, *Bull. Environ. Contam. Toxicol.*, 23, 465, 1978.
39. Rao, K. R., *Pentachlorophenol: Chemistry, Pharmacology and Environmental Toxicology*, Plenum Press, New York, 1978.
40. Hanumante, M. M. and Kulkarni, S. S., Acute toxicity of two molluscicides, mercuric chloride and pentachlorophenol to a freshwater fish *(Channa gachua), Bull. Environ. Contam. Toxicol.*, 23, 725, 1979.
41. Boyle, T. P., Robinson-Wilson, E. F., Petty, J. D., and Weber, W., Degradation of pentachlorophenol in simulated lentic environment, *Bull. Environ. Contam. Toxicol.*, 24, 177, 1980.
42. Phipps, G. L., Holcombe, G. W., and Fiandt, J. T., Acute toxicity of phenol and substituted phenols to the fathead minnow, *Bull. Environ. Contam. Toxicol.*, 26, 585, 1981.
43. Aderson, A. C., Abdelghani, A. A., Smith, P. M., Mason, J. W., and Englande, A. J., Jr., The acute toxicity of MSMA to black bass (*Micropterus dolomieu)*, crayfish (*Procambarus* sp.) and channel catfish *(Ictalurus lacustris), Bull. Environ. Contam. Toxicol.*, 14, 330, 1975.
44. Gray, C. and Knowles, C. O., Metabolic fate and tissue residues of propamocarb in bluegills and channel catfish, *Chemosphere*, 10, 469, 1981.
45. Hermanutz, R. O., Mueller, L. H., and Kempfert, K. D., Captan toxicity to fathead minnows (*Pimephales promelas)*, bluegills (*Lepomis macrochirus)*, and brook trout *(Salvelinus fontinalis), J. Fish. Res. Board Can.*, 30, 1811, 1973.
46. Hildebrand, L. D., Sullivan, D. S., and Sullivan, T. S., Experimental studies of rainbow trout populations exposed to field applications of Roundup® herbicide, *Arch. Environ. Contam. Toxicol.*, 11, 93, 1982.
47. Ingham, B. and Gallo, M. A., Effect of Asulam in wildlife species acute toxicity to birds and fish, *Bull. Environ. Contam. Toxicol.*, 13, 194, 1975.
48. Sanders, H. O. and Hunn, J. B., Toxicity, bioconcentration, and depuration of the herbicide Bolero 8EC in freshwater invertebrates and fish, *Bull. Jpn. Soc. Sci. Fish.*, 48, 1139, 1982.
49. Olson, L. E., Allen, J. L., and Mauck, W. L., Dinitramine: residues in and toxicity to freshwater fish, *J. Agric. Food Chem.*, 23, 437, 1975.
50. Crosby, D. G. and Tucker, R. K., Toxicity of aquatic herbicides to *Daphnia magna, Science*, 154, 289, 1966.
51. Zitko, V., McLeese, D. W., Metcalfe, C. D., and Carson, W. G., Toxicity of permethrin, decamethrin, and related pyrethroids to salmon and lobster, *Bull. Environ. Contam. Toxicol.*, 21, 338, 1979.
52. Kanazawa, J., Measurement of the bioconcentration factors of pesticides by freshwater fish and their correlation with physicochemical properties of acute toxicities, *Pestic. Sci.*, 12, 417, 1981.
53. Sprague, J. B., Measurement of pollutant toxicity of fish. III. Sublethal effects and "SAFE" concentrations, *Water Res.*, 5, 245, 1971.
54. Hart, W. B., Weston, R. F., and DeMann, J. G., An apparatus for oxygenating test solutions in which fish are used as test animals for evaluating toxicity, *Trans. Am. Fish. Soc.*, 75, 228, 1948.
55. Burdick, G. E., Use of bioassays in determining levels of toxic wastes harmful to aquatic organisms, *Am. Fish. Soc.*, Spec. Publ. No. 4, 3.
56. Report on Fish Toxicity Testing Procedures, Tech. Pap. 24, European Inland Fisheries Advisory Commission, 1975, 25, in Lloyd, R., The use of the concentration-response relationship in assessing acute fish toxicity data, in *Analyzing the Hazard Evaluation Process*, Dickson, K. L., Maki, A. W., and Cairns, J., Jr., Eds., American Fisheries Society, Washington, D.C., 1979, 58.
57. Lloyd, R., The use of the concentration-response relationship in assessing acute fish toxicity data, in *Analyzing the Hazard Evaluation Process*, Dickson, K. L., Maki, A. W., and Cairns, J., Jr., Eds., American Fisheries Society, Washington, D.C., 1979, 58.
58. International Joint Commission, New and Revised Great Lakes Water Quality Objectives, Windsor, Ontario, 1977, 1.

59. Stephan, C. E. and Mount, D. I., Use of toxicity tests with fish in water pollution control, in *Biological Methods for the Assessment of Water Quality*, ASTM STP 528, American Society for Testing and Materials, Philadelphia, 1973, 164.

60. Mount, D. I. and Stephan, C. E., A method for establishing acceptable toxicant limits for fish — malathion and the butoxyethanol ester of 2,4-D, *Trans. Am. Fish. Soc.*, 96, 185, 1967.

61. Eaton, J. G., Recent developments in the use of laboratory bioassays to determine "SAFE" levels of toxicants for fish, in *Bioassay Techniques and Environmental Chemistry*, Ann Arbor Science, Ann Arbor, Mich., 1973, 107.

62. Kenaga, E. F., Aquatic test organisms and methods useful for assessment of chronic toxicity of chemicals, in *Analyzing the Hazard Evaluation Process*, Dickson, K. L., Maki, A. W., and Cairns, J., Jr., Eds., American Fisheries Society, Washington, D.C., 1979, 101.

63. Eaton, J. G., Chronic malathion toxicity to the bluegill *(Lepomis macrochirus, Rafinesque)*, *Water Res.*, 2, 215, 1968.

64. Johnson, W. W. and Julin, A. M., Acute toxicity of toxaphene to fathead minnows, channel catfish, and bluegills, EPA-600/3-80-005, U.S. Environmental Protection Agency, Duluth, Minn., 1980.

65. Mauck, W. L., Mehrle, P. M., and Mayer, F. L., Effects of the polychlorinated biphenyl Aroclor 1245 on growth, survival, and bone development in brook trout *(Salvelinus fontinalis)*, *J. Fish. Res. Board Can.*, 35, 1084, 1978.

66. McKim, J. M., Evaluation of tests with early life stages of fish for predicting long-term toxicity, *J. Fish. Res. Board Can.*, 34, 1148, 1977.

67. Maki, A. W., Correlations between *Daphnia magna* and fathead minnow (*Pimephales promelas*) chronic toxicity values for several classes of test substances, *J. Fish. Res. Board Can.*, 36, 411, 1979.

68. Pickering, Q. H., Henderson, C., and Lemke, A. E., The toxicity of organic phosphorus insecticides to different species of warmwater fishes, *Trans. Am. Fish. Soc.*, 91, 175, 1962.

69. Macek, K. J. and McAllister, W. A., Insecticide susceptibility of some common fish family representatives, *Trans. Am. Fish. Soc.*, 99, 20, 1970.

70. Post, G. and Schroeder, T. R., The toxicity of four insecticides to four salmonid species, *Bull. Environ. Contam. Toxicol.*, 6, 144, 1971.

71. Eisler, R., Acute toxicities of organochlorine and organophosphorus insecticides to estuarine fishes, Tech. Pap. No. 46, Bureau of Sport Fisheries and Wildlife, Fish and Wildlife Service, U.S. Department of the Interior, Washington, D.C., 1970.

72. Buccafusco, R. J., Ells, S. J., and LeBlanc, G. A., Acute toxicity of priority pollutants to bluegill *(Lepomis macrochirus)*, *Bull. Environ. Contam. Toxicol.*, 26, 446, 1981.

73. Ambient Water Quality Criteria for DDT, EPA 440/5-80-038, Office of Water Regulations and Standards Criteria and Standards Division, Environmental Protection Agency, Washington, D.C., 1980.

74. Marking, L. L., A method for rating chemicals for potency against fish and other organisms, in Investigations in Fish Control 36, Fish and Wildlife Service, U.S. Department of the Interior, Washington, D.C., 1970.

75. Marking, L. L. and Willford, W. A., Comparative toxicity of 29 nitrosalicylanilides and related compounds to eight species of fish, in Investigations in Fish Control 37, Fish and Wildlife Service, U.S. Department of the Interior, Washington, D.C., 1970.

76. Iyatomi, K., Tamura, T., Itazawa, Y., Hanyu, I., and Sugiura, S., Toxicity of endrin to fish, *Prog. Fish-Culturist*, 155, 1958.

77. Bills, T. D. and Marking, L. L., Toxicity of 3-trifluoromethyl-4-nitrophenol (TFM), 2′,5-dichloro-4′-nitrosalicylanilide (Bayer 73), and a 98:2 mixture to fingerlings of seven fish species and to eggs and fry of coho salmon, in Investigations in Fish Control 69, Fish and Wildlife Service, U.S. Department of the Interior, Washington, D.C., 1976.

78. Olson, L. E. and Marking, L. L., Toxicity of four toxicants to green eggs of salmonids, *Prog. Fish-Culturist*, 37, 143, 1975.

79. Folmar, L. C., Sanders, H. O., and Julin, A. M., Toxicity of the herbicide glyphosate and several of its formulations to fish and aquatic invertebrates, *Arch. Environ. Contam. Toxicol.*, 8, 269, 1979.

80. Marking, L. L., Bills, T. D., and Chandler, J. H., Toxicity of the lampricide 3-trifluoromethyl-4-nitrophenol (TFM) to nontarget fish in flow-through tests, in Investigations in Fish Control 61, Fish and Wildlife Service, U.S. Department of the Interior, Washington, D.C., 1975.

81. Bull, C. J. and McInerney, J. E., Behavior of juvenile coho salmon (*Oncorhynchus kisutch*) exposed to sumithion (*Fenitrothion*), an organophosphate insecticide, *J. Fish. Res. Board Can.*, 31, 1867, 1974.

82. Schimmel, S. C., Hansen, D. J., and Forester, J., Effects of Aroclor® 1254 on laboratory-reared embryos and fry of sheepshead minnows (*Cyprinodon variegatus*), *Trans. Am. Fish. Soc.*, 103, 582, 1974.

83. Klaverkamp, J. F., Duangsawasdi, M., Macdonald, W. A., and Majewski, H. S., An evaluation of fenitrothion toxicity in four life stages of rainbow trout, *Salmo gairdneri,* in *Aquatic Toxicology and Hazard Evaluation,* ASTM STP 634, Mayer, F. L. and Hamelink, J. L., Eds., American Society for Testing and Materials, Philadelphia, 1977, 231.

84. Buhler, D. R., Rasmusson, M. E., and Shanks, W. E., Chronic oral DDT toxicity in juvenile coho and chinook salmon, *Toxicol. Appl. Pharmacol.,* 14, 535, 1969.

85. Buhler, D. R. and Shanks, W. E., Influence of body weight on chronic oral DDT toxicity in coho salmon, *J. Fish. Res. Board Can.,* 27, 347, 1970.

86. Lee, J. H., Sylvester, J. R., and Nash, C. E., Effects of mirex and methoxychlor on juvenile and adult striped mullet, *Mugil cephalus* L., *Bull. Environ. Contam. Toxicol.,* 14, 180, 1975.

87. Kumaraguru, A. K. and Beamish, F. W. H., Lethal toxicity of permethrin (NRDC-143) to rainbow trout, *Salmo gairdneri,* in relation to body weight and water temperature, *Water Res.,* 15, 503, 1981.

88. Gupta, S., Verma, S. R., and Saxena, P. K., Toxicity of phenolic compounds in relation to the size of a freshwater fish, *Notopterus notopterus* (Pallas), *Ecotoxicol. Environ. Saf.,* 6, 433, 1982.

89. Norris, L. A. and Miller, R. A., The toxicity of 2,3,7,8-tetrachlorodibenzo-p-dioxin (TCDD) in guppies *(Poecilia reticulatus Peters), Bull. Environ. Contam. Toxicol.,* 12, 76, 1974.

90. Adelman, I. R., Smith, L. L., and Siesennop, G. D., Effect of size or age of goldfish and fathead minnows on use of pentachlorophenol as a reference toxicant, *Water Res.,* 10, 685, 1976.

91. Jarvinen, A. W., Hoffman, M. J., and Thorslund, T. W., Long-term toxic effects of DDT food and water exposure on fathead minnows *(Pimephales promelas), J. Fish. Res. Board Can.,* 34, 2089, 1977.

92. Colgan, P. W., Cross, J. A., and Johansen, P. H., Guppy behavior during exposure to a sub-lethal concentration of phenol, *Bull. Environ. Contam. Toxicol.,* 28, 20, 1982.

93. Marking, L. L., personal communication.

Chapter 2

INFLUENCE OF ENVIRONMENTAL CONDITIONS ON THE TOXICITY OF PESTICIDES TO FISH

I. INTRODUCTION

In nature, the overall effect of a pesticide on the biota is influenced by a number of environmental conditions which may alter the rate of metabolism of the organism, and the state and availability of the toxicant. Alternatively, these factors may get the toxicant out of solution or increase its adsorption to the particulate matter, thereby making it unavailable. As a consequence, the extent of uptake by the organism is altered. Hence, from the early days of toxicity testing, many attempts were made to study the influence of environmental factors on the toxicity of pesticides to the aquatic organisms. In this chapter, the effect of temperature, hydrogen ion concentration, hardness of water, turbidity, and salinity on the bioavailability and toxicity of pesticides to fish is considered. Also, the relevance of laboratory data to the field conditions is discussed.

II. INFLUENCE OF ENVIRONMENTAL FACTORS

A. Temperature

Among all the environmental factors, the effect of temperature on the toxicity of chemicals is studied most often. Temperature acts directly or indirectly through its influence on other factors like enzyme activity, metabolic rate, and certain other environmental factors such as dissolved oxygen. Aquatic organisms may be exposed to greater amounts of the toxicant at higher temperatures, because of increased diffusion and active uptake. To a certain extent, however, accelerated uptake of the toxicant and the resultant increase in toxicity are countered by elevated levels of detoxification and metabolization.

Toxicity of DDT decreases with increase in temperature and vice versa.[1] Delayed mortality of the Atlantic salmon parr at the onset of the cold conditions in autumn was reported.[2] Following the spraying of New Brunswick forests with DDT in June, heavy mortality of the parr and young salmon was noticed when the water temperature reached 5°C and below.[3] Sanders and Cope[4] reported decreased toxicity of DDT to cladocerans at higher temperatures (48-h EC 50 to *Simocephalus serrulatus* at 10 and 27°C was 1.7 and 4.3 $\mu g/\ell$). When DDT-exposed brook trout were stressed by starvation, the beginning of appreciable mortality coincided with the onset of colder temperature of water.[5] The mortality of DDT-exposed fish in this case was perhaps caused by an interaction of DDT residues with a combination of environmental stresses like starvation, decreasing water temperature and increasing physiological stress associated with spawning. Methoxychlor, closely related to DDT structurally, also behaves like DDT, with decreased toxicity to fish at higher temperature; the 96-h LC 50s to rainbow trout and bluegill were, respectively, 30 and 42 $\mu g/\ell$ at 1.6°C and 62 and 75 $\mu g/\ell$ at 12.7°C.[6]

DDT altered the preferred temperatures of fish. Ogilvie and Anderson[7] showed that DDT-exposed Atlantic salmon selected a temperature different from that of the controls. The preferred temperature depended on the concentrations to which the fish were exposed; those that were exposed to lower concentrations consistently selected a lower temperature than that of the controls, whereas those that were exposed to higher concentrations selected a temperature higher than that of the controls. This shift appeared

Table 1
THE INFLUENCE OF TEMPERATURE ON THE
TOXICITY OF ENDOSULFAN (μg/ℓ) TO FISH
AND INVERTEBRATES

Species	Temperature (°C)	96-h LC 50	120-h LC 50
Rainbow trout	1.5	0.8	0.7
	10	0.3	0.3
Western white sucker	10	3.5	2.5
	19	3	2.8
Daphnia magna	10	178[a]	47.5
	19	68[a]	53.5
Damselfly naiad	8	71.8	62
	19	107	75

[a]   24-h LC 50.

Reprinted from Schoettger, R. A., Toxicology of thiodan in several fish and aquatic invertebrates, Investigations in Fish Control, Bureau of Sport Fisheries and Wildlife, U.S. Department of the Interior, Washington, D.C., 1970. With permission.

more marked in those fish initially acclimated to higher temperatures (17°C) than those acclimated to lower temperatures (8°C).[7] Similar results were reported in the case of rainbow trout.[8] A single, 24-hr exposure of underyearling Atlantic salmon to 50 μg/ℓ elevated the preferred temperature, and this shift persisted for at least a month after the fish were transferred to clean water.[9] Exposure of brook trout to 20 to 30 μg/ℓ concentration of *p,p'*-DDT, *o,p'*-DDT, *p,p'*- and *o,p'*-DDD, *p,p'*-DDA, *p,p'*-methocychlor, *p,p'*-Cl DDT, and *p,p'*- and *o,p'*-DDE resulted in the selection of lower temperature, except in the case of the last two compounds. A 24-hr exposure to DDT caused a selection of temperatures 4 to 4.5°C lower than the normal, during the entire period of experimentation (9 days); with methoxychlor, there was an initial 5°C lowering of the selected temperature, but this was progressively reversed, and by the 5th day, the selected temperature was not significantly different from that of the controls. This reversal was probably possible because methoxychlor, unlike DDT, was biodegradable.[10]

In many other investigations on the effect of temperature on the acute toxicity of other pesticides, a direct relation between the two — unlike in the case of DDT and methoxychlor — was reported. Macek et al.[6] studied the toxicity of 10 pesticides to the rainbow trout and 11 pesticides to the bluegills at different temperatures (1.6, 7.2, and 12.7°C in the case of the former, and 12.7, 18.3, and 23.8°C with the latter) and found that the toxicity increased with increasing temperatures. Toxicity of endrin to goldfish increased with temperature; 48-h LC 50 at 4°C was 0.14; at 17 to 18°C, 0.004 to 0.008, and at 27 to 28°C, 0.002 (mg/ℓ). The catfish, *Heteropneustes fossilis,* was more tolerant to endosulfan in the colder months of the year than in the warmer months.[11] Schoettger[12] also noted increased toxicity of endosulfan to the rainbow trout and *Daphnia* with increase in temperature; however, its toxicity to the damsel-fly, *Ischnura* sp., decreased with increasing temperature (Table 1). The 24-h LC 50 value of trifluralin to bluegills at 29°C was 130 times as high as at 7°C.[13]

Toxicity of permethrin to the rainbow trout was inversely related to the temperature, the 96-h LC 50 being 0.62 μg/ℓ at 5°C and 6.4 μg/ℓ at 20°C.[14] The toxicity of natural pyrethrum and some synthetic phyrethroids (dimethrin, RU-11679, resmethrin) was higher at 12°C than at 17°C; dimethrin and resmethrin were more toxic at 17°C than at 22°C but the toxicity of RU-11679, and that of the natural pyrethrums was not affected by the latter temperature.[15] On the other hand, α-trans allethrin was more toxic at higher temperature than at lower temperature. Cold-water species were more sensitive to the action of both natural and synthetic pyrethroids than warm-water fish.[15]

Diazinon was more toxic to fish at higher temperatures.[1] Likewise, the toxicity of fenitrothion and acephate to rainbow trout increased with increasing temperature.[16] As in the case of DDT and some pyrethroids, toxicity of phenol to rainbow trout was less at higher temperatures. Also, at higher temperatures, all the recorded deaths were observed only in the first 24 hr, whereas at 6°C, deaths were recorded until 36 hr.[17] The exposure time necessary to effect 100% mortality of the unwanted fish by using rotenone, was dependent on the water temperature. Considerably longer exposure was needed at lower temperatures, irrespective of the concentration.[18] Many more examples of the effect of temperature on the toxicity of pesticides are mentioned in the compilation of the data on toxicity tests conducted at the Columbia National Fisheries Research Laboratory (CNFRL), Missouri.[19]

A higher uptake of DDT by fish at higher temperatures was recorded, as a result of the increased metabolic rate.[20] Similarly, the uptake of polychlorinated biphenyls (PCBs) by rainbow trout, fathead minnow, green sunfish,[21] and methylmercury by bluntnose minnows[22] was higher at higher temperature.

The higher toxicity of many pesticides at higher temperature can be explained on the basis of increased uptake of the toxicant because of a higher ventilation rate.[6] But the decrease in the toxicity of DDT and methoxychlor is contrary to the normal expectations. Gardner[10] opined that molecules with an ethane carbon configuration attached to o,p' or p,p'-substituted phenyl rings affect the temperature preference of fish. Jackson and Gardner,[23] while reporting on the in vitro inhibition of Mg^{2+}-activated ATPases by DDT, noted that DDT and several of its analogs increased the energy of activation and the frequency of the enzyme-catalyzed reaction, resulting in a negative temperature coefficient of inhibition.

Cairns et al.[24] reviewed the effects of temperature on the toxicity of chemicals to the aquatic organisms, and noted, "Although rather extensive bibliographies give the impression that there is a vast amount of literature on the effects of temperature on aquatic organisms, when one tries to apply this information to specific interactions, often very little of the evidence is available."

B. Hydrogen Ion Concentration (pH)

Weber[25] presented an excellent review on the various physicochemical properties of pesticides vis-a-vis pH. Hydrolysis, which is a major pathway of degradation of many compounds in the aquatic system, is dependent on pH. For instance, the hydrolysis of S-triazines, especially that of the chloro-substituted compounds, is rapid under highly acidic or alkaline conditions. Volatilization, which contributes to the loss of chemicals from aquatic systems, is also dependent on vapor pressure and pH. Weber also discussed the relationship between the dissociated and undissociated forms of ionic compounds.

With few exceptions, the toxicity of organochlorine (OC) compounds is not affected by changes in the hydrogen ion concentration.[26-28] Malathion is toxic below pH 7, and loses its toxicity in alkaline pH, as it undergoes hydrolysis at pH 7.[29] Phosmet was 33 times less toxic at pH 8.5 than at pH 6.5.[19] The majority of the OC and organophosphate (OP) compounds are nonionic and their activity is not influenced by pH. Among OP compounds, pH influences the toxicity to the extent it controls the hydrolysis of the compound. On the contrary, ionic compounds are greatly influenced by the prevailing pH, because the extent of dissociation of an ionic compound is dependent on the pH. The undissociated form is toxic, whereas the dissociated form (ion) does not penetrate the biological membranes and hence is not so toxic.[30] At a particular pH, more of the undissociated form of a compound may be in existence, contributing to the toxic effect of the compound. The dissociation constant (pKa) of a compound is the pH at which 50% of it is in the dissociated state. Above or below this pH, depending on the acidic or basic nature of the compound, more or less of the compound will be in the dissociated state, which easily explains why pH influences the toxicity of ionic compounds.

The toxicity of pentachlorophenol (PCP) to fish was reduced by an increase in the pH from 6.5 to 7.5.[31] Similarly, the toxic effects of 2,4-dichlorophenoxyacetic acid (2,4-D) were reduced when the pH was raised by the addition of NaOH. The percent of fathead minnows that survived a particular concentration of 2,4-D increased with increasing pH. At a concentration of 7.43 mg/ℓ, 60% of the fish survived at pH 7.57, and 100% at pH 8.25. At this concentration, 28% survived in 192 hr at a pH 7.6, whereas 100% survived at pH 9.8. Normal schooling behavior was completely disrupted and the equilibrium was lost after 24-hr exposure to 7.43 mg/ℓ at pH 7.6, whereas neither character was affected even after 192-hr exposure to this concentration, at a higher pH (8.68 or 9.08). The reduction in the toxicity of 2,4-D is due to a greater amount of dissociation in higher pH. The pKa of 2,4-D is 7.8. These findings of Holcombe et al.[30] confirm the earlier results of Crandall and Goodnight.[32] Evidently, alkaline waters can tolerate slightly higher levels of 2,4-D and such other weak acids, which have low pKa values.

The toxicity of 2,4-D acid, butyl ester, and dimethyl amine salt formulations at pH 8.5 was half of that at pH 6.5, but the dodecyl/tetradecyl amine formulation was nearly four times more toxic at pH 8.5 than at 6.5 to the fathead minnows.[19] Higher toxicity of phenols (2,4,6-trichlorophenol) to fish under acidic conditions, reported by Virtanen and Hattula,[33] is in conformity with the above explanation of the dissociation of ionic compounds.

Kaila and Saarikoski[34] reported that there was a decrease in the toxicity of PCP and 2,3,6-trichlorophenol to the crayfish, *Astacus fluviatilis,* when the pH was increased from 6.5 to 7.5; however, this decrease was not as much as that predicted from pKa values. Hence, they postulated that the phenate ion is also toxic. Saarikoski and Viluksela[35] studied the toxicity of six chloro-, bromo-, and nitro-substituted phenols to the guppy, in the pH range of 5 to 8. The toxicity of acidic phenols decreased as the pH increased, whereas that of 4-chlorophenol did not change, as it remained in the unionized state in the pH range studied. Once again, the decrease in the toxicity was smaller than what it would be if the phenate ion were to be nontoxic. They reiterated that the phenate ion also contributes to the toxicity, which diminished with increasing pH.

The 96-h LC 50 of bromoxynil to goldorfe was 0.2 mg/ℓ at pH 6.2, 2 mg/ℓ at pH 7.2, and 20 mg/ℓ at pH 8.2.[36] This finding is in agreement with the discussion on the pKa values of ionic compounds by Weber.[25] Dinoseb was also more toxic to cutthroat trout and lake trout in acidic pH (6.5) than in alkaline pH (8.5).[37]

For a few other compounds, the toxicity was reported to increase with pH. Mexacarbate, a carbamate compound, was 38 times more toxic to bluegills at pH 9.5 than at 7.5, the 96-h LC 50 being 22.9 mg/ℓ at pH 7.5 and 0.6 at pH 9.5. It is difficult to explain this result on the basis of pKa, since the estimated pKa of mexacarbate is 2.6, suggesting that much of the compound was in its undissociated state at both pH 6.5 and 9.5. Hence, the authors suggested that the increased toxicity at higher pH is not related to the greater absorption of the unionized compound, but possibly to the greater toxicity of the degradation product.[38] The toxicity of picloram to cutthroat trout and lake trout increased with a rise in pH from 6.5 to 8.5, and so was the case with a formulation of glyphosate[39] and piscicide, GD-174.[40] Many other examples on the pH-related changes in the toxicity of herbicides can be found in the compilation by Johnson and Finley.[19]

While the activity of the synthetic pyrethroids was not altered by a change in the pH, the biological activity of natural pyrethrums was influenced by pH, its toxicity being higher at lower pH (96-h LC 50 to bluegills was 87 $\mu g/l$ at pH 9.5 and 41 $\mu g/l$ at pH 6.5).[15]

Grohmann and Sobhani[36] suggested that LC 50 of ionic compounds should be determined separately for the acid-base forms (or more appropriately the undissociated and the dissociated forms), because of the influence of pH on the toxicity of ionic compounds.

C. Hardness

As is the case with pH, hardness of water does not seem to influence the toxicity of OC and OP compounds.[26,27] The toxicity of phenol, aniline, and chlorobenzene to the various developmental stages of largemouth bass, channel catfish, goldfish, and rainbow trout at two levels of hardness of water (50 and 200 mg/l, as $CaCo_3$) was studied. Phenol was less toxic in soft water (50 mg/l), whereas the toxicity of aniline and chlorobenzene was not influenced by changes in the hardness of the test medium.[41] The toxicity of three formulations of 2,4-D, three formulations of endothall, fenoprop, PCP, and dichlobenil was not affected by hardness.[42] Hydrothal-191 was more toxic to golden shiners in hard water than in soft water.[43] The lampricides TFM (3-trifluoromethyl-4-nitrophenol) and Bayer® 73 were more toxic to insect larvae in soft water.[44,45] Hardness of water had no effect on the toxicity of mexacarbate[38] or carbaryl.[19] While water hardness had no effect on the toxicity of synthetic pyrethroids, natural pyrethrum was more toxic in hard water.[15] Diquat was less toxic in hard water.[15]

Hardness plays an important role in determining the toxicity of metals. With a few exceptions, hardness does not seem to have great influence on the action of pesticides. Investigating the suggestions of earlier workers that hardness may play a role in determining the toxicity of pesticides, Grohmann and Sobhani[36] concluded that pH, rather than the hardness of water, controls the toxicity of bromoxynil to fish. Johnson and Finley,[19] in their summary on toxicity tests conducted at CNFRL, commented on the influence of hardness or pH or both on the toxicity of 30 compounds. Out of 19 compounds on which the influence of hardness was tested, hardness had no influence in 17 cases; in the other 2 cases pH also was found to control the toxicity. Out of six compounds on which the influence of both pH and hardness was tested, pH and not hardness was the controlling factor in five cases; the toxicity of the sixth compound was altered by both pH and hardness. Hence, it appears that very rarely hardness influences the toxicity of pesticides. In many instances in the past, where hardness was claimed to have increased or decreased the toxicity of a particular compound, it was in all probability the pH that effected this change. Perhaps it would be worthwhile to have a fresh look at this problem, by testing the influence of pH and hardness separately and together, in the case of all those pesticides supposed to have been influenced by hardness too.

D. Turbidity

Suspended particles, both living and nonliving, alter the toxicity of a compound by influencing the extent of bioavailability of a compound. In Chapter 1, Volume I, the various factors that influence the availability of a compound in true solution and the factors that favor its adsorption to particulate matter have been reviewed. Useful reviews on this topic are those of Weber[25] and Bailey and White.[46]

An application of DDT at a rate of 0.09 mg/l for 16 min in the Saskatchewan River eliminated the blackfly larvae over a stretch of 98 mi. A similar treatment in other areas eliminated the insect larvae in only a stretch of 9 mi. The difference was attrib-

Table 2

EFFECTS OF TYPE AND AMOUNT OF SOIL ON
THE TOXICITY OF PARATHION, OR
PARATHION AND ATRAZINE TOGETHER[a]

| | 24-hr mortality (%) of larvae in water with | | | | |
| | | Sand | | Loam | |
Pesticide conc (mg/ℓ)	No soil	1 g	5 g	1 g	5 g
Atrazine (10)	0	0	0	0	0
Parathion (0.015)	24 ± 7	16 ± 7	2 ± 2	7 ± 0	0
Atrazine (10) + Parathion (0.015)	62 ± 8	42 ± 10	2 ± 2	22 ± 4	0
Atrazine (20)	0		0		0
Parathion (0.30)	93 ± 6		62 ± 8		0
Atrazine (20) + Parathion (0.30)	98 ± 4		76 ± 4		38 ± 10

[a]   Data from Liang, T. T. and Lichtenstein, E. P., *Science,* 186, 1128, 1974. Copyright 1974, American Association for the Advancement of Science; reprinted with permission of AAAS.

uted to the high turbidity in the former case, with a suspended particle load of 551 mg/ ℓ. It was explained that in the presence of suspended particles, most of the DDT was adsorbed to the particulate matter, and because of the greater consumption of suspended particles by *Simulium* larvae, a better control ensued in areas of high turbidity.[47]

To examine the influence of suspended particles on the toxicity of xenobiotic chemicals, Brungs and Bailey[48] suspended 50 mg/ℓ of montmorillonite clay, Brookston silty clay loam, or activated carbon (particle size of clay or carbon, 1 to 2 μm) and added endrin to the system. The LC 50 values calculated with the clays were similar to those of the controls, but in the presence of activated carbon, the LC 50 value was approximately 50 times that of the controls. In the latter, too, the LC 50 was comparable to that of the controls. However, the endrin concentration in true solution alone, and not that adsorbed to the carbon particles, was taken into consideration. The presence of suspended particles reduced the mortality of mosquito larvae caused by parathion, or parathion and atrazine together (Table 2).[49] Presence of clay particles in the test tanks greatly reduced the toxicity of lindane to a number of test fish.[50]

Although Brungs and Bailey,[48] quoting Ferguson et al.[51] suggested that chemicals adsorbed to suspended particles were not bioavailable, the work of Zitko,[52] Halter and Johnson,[53] Nau-Ritter and Wurster,[54] Nau-Ritter et al.,[55] and several others showed that chemicals adsorbed to sediments and suspended particles leach into water and as such are absorbed by the aquatic organisms. Such a desorption is a function of the water solubility of the compound and the levels of organic matter. Lower chlorinated biphenyls partitioned into water to a greater extent than the poorly water-soluble, higher chlorinated biphenyls.[56] In eutrophic lakes, in the presence of greater amount of organic material, less of the adsorbed DDT was available for absorption, whereas in oligotrophic lakes, the presence of more suspended matter meant availability of more dieldrin for absorption.[57]

The influence of suspended matter on the acute toxicity of pesticides to fish has not been investigated to a great extent. This problem deserves a closer inspection, especially with hydrophobic compounds that have been tested near or above their saturation limits in water. When a compound is employed above its limit of water solubility, the amount in excess of the saturation limits will be present in precipitation, or adsorbed to the walls of the test containers. The suspended particles also adsorb the amount that is in excess of the water solubility. Traditionally, the LC 50 values have been calculated on the basis of the quantity of the compound that had been introduced into the test tanks, or in a few instances, on the total amount present in the system that could be extracted from water and quantified. (It should be emphasized that a solvent extracts not only the material in true solution, but also that in suspension and in adsorption to the particulate matter.) In either case, the LC 50 values of the hydrophobic compounds reported earlier exceeded the water saturation concentration of those compounds. Herzel and Murty[58] opined that LC 50 values expressed in excess of the water solubility limit of a compound are unacceptable, since all the material in the tank is not truly bioavailable. The finding of Brungs and Bailey,[48] that out of 26 μg endrin per liter present in the test tank, only 0.62 μg/l was really toxic to half of the fathead minnows in 96-hr, supports this contention.

E. Influence of Other Environmental Factors on the Toxicity of Pesticides to Fish

There have been but a few studies on the alteration of pesticide toxicity in saline waters. Brown et al.[59] reported that the toxicity of phenol to rainbow trout increased with salinity. Murphy[60] reported that young mosquitofish accumulated significantly lesser amounts of DDT, DDE, and DDD in 15% saline waters than in freshwater. Similarly, juvenile Atlantic salmon accumulated higher quantities of 2,2',4,5,5'-pentachlorobiphenyl in freshwater than in sea water.[61]

Not much work seems to have been carried out on the influence of low oxygen on the toxicity of pesticides to fish. The toxicity of zinc to bluegills[62] and hydrogen sulfide to goldfish[63] was higher when the oxygen concentration was lower.

III. EXTRAPOLATION OF LABORATORY DATA TO THE FIELD CONDITIONS

Since the toxicity of pesticides to fish is altered by many environmental factors, it would be interesting to ponder over the dependability of the laboratory-generated data for purposes of environmental hazard evaluation.

In nature, adsorption to plants and organic matter rich in suspended material makes the toxicant less bioavailable. Even in one of the earliest reports on the toxicity of DDT to fish, it was mentioned that the presence of aquatic plants in the test containers reduced the toxicity of DDT to goldfish.[64] Toxicity of lindane to freshwater fish was much less in natural ponds than in the aquarium tanks.[50] Humic acid, which is the alkali-soluble, acid-insoluble fraction of the humus, solubilizes DDT in water. Hence, in nature more DDT can be present in solution and be bioavailable than under laboratory conditions.[65] Hamelink and Waybrant[66] emphasized that one problem in extrapolating laboratory experiments to the field situations is that the laboratory models cannot simulate natural conditions because of their high sediment to water ratio and the high air to water interphase ratio. Mount[67] also stressed that laboratory data cannot be extrapolated completely to predict natural conditions or to protect the natural communities adequately. Laboratory data can predict community effects well, if one is concerned with the gross effects, but they cannot predict well the refined effects. Normal variations and fluctuations that occur in the community structure also prevent the testing of the adequacy of the laboratory data until the predicted change is great enough to exceed the natural variation. Mount pointed out that the usual tendency is to underestimate or overestimate the possible toxicity of a compound.

Freshwater fish bioconcentrate DDT under field conditions much more than in the laboratory experiments, which may be due to the difference in the physical form of the toxicant and additional trophic levels involved in the field.[68] The oxidative hydroxylation of carbofuran in the liver of wild *Trichogaster pectoralis* was comparatively lower than in the experimental fish.[69] Rainbow trout took up lesser quantities of ^{14}C-triphenylphosphate from river water than from dechlorinated Winnipeg city tap water. The difference could not be attributed either to degradation or to sorption to the suspended material. Water type seems to have caused this difference, and residues in the river water were less available to the fish.[70] Similar results were obtained in experiments with 2-ethyldiphenylphosphate[71] and terbutryn and fluridone.[72] Analyzing the reasons for the lesser accumulation of methoxychlor and other compounds by fish and other aquatic organisms in natural water than in dechlorinated city water, Lockhart et al.[73] suggested that adsorption to suspended materials may make the pollutants less available for bioaccumulation. Methoxychlor adsorbed on the celite particles was less toxic than an emulsifiable concentration.

Certain species of fish like *Barbus* which are very susceptible to endosulfan in laboratory experiments were rarely found dead in nature after aerial application of endosulfan for the control of tsetse fly.[74] Consequent upon the exposure to PCB, size-specific mortality (with the surviving lake trout fry being larger) was observed in the fry collected from Lake Michigan but not in hatchery-reared lake trout fry.[75] Significant differences in the extent of accumulation of certain contaminants released from dredged sediments were evident between hatchery-reared fish and those of Great Lakes origin. The uptake by the latter was often significantly higher.[76]

The half-life of fluridone in sediment, under laboratory conditions, was 12 months and under field conditions 17 weeks.[77] Owing to changes induced in the toxicity and uptake of xenobiotic chemicals by previous and simultaneous exposure to other contaminants in the field, extrapolation of the laboratory data to the field is difficult.[78] Greater sensitivity of fish and invertebrates to the chemicals in laboratory tests led Chandler and Marking to suggest that the ultimate safety of chemicals should be evaluated under the field conditions.[79]

The ultimate effects of a pesticide in the environment are dependent on many interrelated phenomena. First and foremost among these factors is the chemistry of the molecule itself; second, the *in situ* properties (physicochemical and biological) of the water body to which the laboratory data have to be extrapolated; and last, the properties and responses of the organism itself influence the extent of toxicity of a compound and its environmental behavior.[80]

Although the above discussion may suggest that toxic effects are greatly altered under field conditions, and no reasonable extrapolation of laboratory data to the field is possible, such an extrapolation is possible when all the above factors are taken into consideration.

IV. CONCLUSIONS

Toxicity testing is undertaken in the laboratory with a view to assessing the toxicity of a compound to the biota. Modification of the toxicity of a compound by environmental factors like temperature, hydrogen ion concentration, load and nature of suspended particles, salinity, etc. renders the extrapolation of the laboratory data to field conditions difficult. In general, rise in ambient temperature brings about an increase in the toxicity of a compound, owing to increased metabolic rate of the organism, and hence increased uptake of a xenobiotic chemical. There are some exceptions to this, the most notable being DDT and its analogs. pH exerts its influence on ionic compounds, the ratio of undissociated to dissociated form determining the toxicity of a compound at any pH. The influence of hardness on the toxicity of pesticides seems limited. Increased turbidity generally renders the pesticide less toxic, owing to increased adsorption; however, increased load of suspended particles may prove more harmful to the detrivores. There is an urgent need to evaluate the role of suspended particles in reducing the bioavailability of hydrophobic compounds, especially when the compound is employed in toxicity tests at an amount greater than its water solubility. The extrapolation of the laboratory data to the field conditions has to be undertaken with great caution. Yet, taking into consideration the chemistry of the molecule, the physicochemical properties of the aquatic bodies, and the responses of the biota, such an extrapolation is reasonably possible.

REFERENCES

1. Cope, O. B., Sport fishery investigations. Laboratory studies and toxicology, *U.S. Bur. Sport Fish. Wildl. Serv. Circ.*, 226, 51, 1965.
2. Delayed losses of salmon follow DDT spraying, Annual Report Fisheries Research Board of Canada, 1960—61, 65; as cited in Elson, P. F., Effects on wild young salmon of spraying DDT over New Brunswick forests, *J. Fish. Res. Board Can.*, 24, 731, 1967.
3. Elson, P. F., Effects on wild young salmon of spraying DDT over New Brunswick forests, *J. Fish. Res. Board Can.*, 24, 731, 1967.
4. Sanders, H. O. and Cope, O. B., Toxicities of several pesticides to two species of cladocerans, *Trans. Am. Fish. Soc.*, 95, 165, 1966.
5. Macek, K. J., Growth and resistance to stress in brook trout fed sublethal levels of DDT, *J. Fish. Res. Board Can.*, 25, 2443, 1968.
6. Macek, K. J., Hutchinson, C., and Cope, O. B., The effects of temperature on the susceptibility of bluegills and rainbow trout to selected pesticides, *Bull. Environ. Contam. Toxicol.*, 4, 174, 1969.
7. Ogilvie, D. M. and Anderson, J. M., Effect of DDT on temperature selection by young Atlantic salmon, *Salmo salar*, *J. Fish. Res. Board Can.*, 22, 503, 1965.
8. Miller, D. L. and Ogilvie, D. M., Temperature selection in brook trout (*Salvelinus fontinalis*) following exposure to DDT, PCB or phenol, *Bull. Environ. Contam. Toxicol.*, 14, 545, 1975.
9. Ogilvie, D. M. and Miller, D. L., Duration of a DDT-induced shift in the selected temperature of Atlantic salmon *(Salmo salar)*, *Bull. Environ. Contam. Toxicol.*, 16, 86, 1976.
10. Gardner, D. R., The effect of some DDT and methoxychlor analogs on temperature selection and lethality in brook trout fingerlings, *Pestic. Biochem. Physiol.*, 2, 437, 1973.
11. Singh, B. B. and Narain, A. S., Acute toxicity of thiodan to catfish (*Heteropneustes fossilis*), *Bull. Environ. Contam. Toxicol.*, 28, 122, 1982.
12. Schoettger, R. A., Toxicology of thiodan in several fish and aquatic invertebrates, in Investigations in Fish Control 35, Fish and Wildlife Service, U.S. Department of the Interior, Washington, D.C., 1970.
13. Cope, O. B., Some responses of fresh-water fish to herbicides, Proc. 18th Annu. Meet. Southern Weed Conf., Dallas, Tex., 1965, 439.
14. Kumaraguru, A. K. and Beamish, F. W. H., Lethal toxicity of permethrin (NRDC-143) to rainbow trout, *Salmo gairdneri*, in relation to body weight and water temperature, *Water Res.*, 15, 503, 1981.
15. Mauck, W. L., Olson, L. E., and Marking, L. L., Toxicity of natural pyrethrins and five pyrethroids to fish, *Arch. Environ. Contam. Toxicol.*, 4, 18, 1976.
16. Duangsawasdi, M. and Klaverkamp, J. F., Acephate and fenitrothion toxicity in rainbow trout: effects of temperature stress and investigations on the sites of action, in Aquatic Toxicology, ASTM STP 667, Marking, L. L. and Kimerle, R. A., Eds., American Society for Testing and Materials, Philadelphia, 1979, 35.
17. Brown, V. M., Jordan, D. H. M., and Tiller, B. A., The effect of temperature on the acute toxicity of phenol to rainbow trout in hard water, *Water Res.*, 1, 587, 1967.
18. Gilderhus, P. A., Exposure times necessary for antimycin and rotenone to eliminate certain fresh-water fish, *J. Fish. Res. Board Can.*, 29, 199, 1972.
19. Johnson, W. W. and Finley, M. T., Handbook of Acute Toxicity of Chemicals to Fish and Aquatic Invertebrates, Resource Publ. 137, Fish and Wildlife Service, U.S. Department of the Interior, Washington, D.C., 1980.

20. Murphy, P. G. and Murphy, J. V., Correlations between respiration and direct uptake of DDT in the mosquitofish *Gambusia affinis, Bull. Environ. Contam. Toxicol.,* 6, 581, 1971.
21. Veith, G. D., DeFoe, D. L., and Bergstedt, B. V., Measuring and estimating the bioconcentration factor of chemicals in fish, *J. Fish Res. Board Can.,* 36, 1040, 1979.
22. Burkett, R. D., The influence of temperature on uptake of methylmercury-203 by bluntnose minnows, *Pimephales notatus* (Rafinesque), *Bull. Environ. Contam. Toxicol.,* 12, 703, 1974.
23. Jackson, D. A. and Gardner, D. R., In vitro effects of DDT analogs on trout brain Mg^{2+}-ATPase, *Pestic. Biochem. Physiol.,* 8, 123, 1978.
24. Cairns, J., Jr., Heath, A. G., and Parker, B. C., The effects of temperature upon the toxicity of chemicals to aquatic organisms, *Hydrobiologia,* 47, 135, 1975.
25. Weber, J. B., Interaction of organic pesticides with particulate matter in aquatic and soil systems, in *Fate of Organic Pesticides in the Aquatic Environment* (Advances in Chemistry Series 111), Faust, S. D., Ed., American Chemical Society, Washington, D.C., 1972, 55.
26. Pickering, Q. H., Henderson, C., and Lemke, A. E., The toxicity of organic phosphorus insecticides to different species of warmwater fishes, *Trans. Am. Fish. Soc.,* 91, 175, 1962.
27. Henderson, C., Pickering, Q. H., and Tarzwell, C. M., Relative toxicity of ten chlorinated hydrocarbon insecticides to four species of fish, *Trans. Am. Fish. Soc.,* 88, 23, 1959.
28. Johnson, W. W. and Julin, A. M., Acute toxicity of toxaphene to fathead minnows, EPA-600/3-80-005, U.S. Environmental Protection Agency, Duluth, Minn., 1980.
29. Bender, M. E., The toxicity of the hydrolysis and breakdown products of malathion to the fathead minnow (*Pimephales promelas,* Rafinesque), *Water Res.,* 3, 571, 1969.
30. Holcombe, G. W., Fiandt, J. T., and Phipps, G. L., Effects of pH increases and sodium chloride additions on the acute toxicity of 2,4-dichlorophenol to the fathead minnow, *Water Res.,* 14, 1073, 1980.
31. Goodnight, C. J., Toxicity of sodium pentachlorophenate and pentachlorophenol to fish, *Ind. Eng. Chem.,* 34, 868, 1942.
32. Crandall, C. A. and Goodnight, C. J., The effect of various factors on the toxicity of sodium pentachlorophenate to fish, *Limnol. Oceanogr.,* 4, 53, 1959.
33. Virtanen, M. T. and Hattula, M. L., The fate of 2,4,6-trichlorophenol in an aquatic continuous-flow system, *Chemosphere,* 11, 641, 1982.
34. Kaila, K. and Saarikoski, J., Toxicity of pentachlorophenol and 2,3,6-trichlorophenol to the crayfish (*Astacus fluviatilis*), *Environ. Pollut.,* 12, 119, 1977.
35. Saarikoski, J. and Viluksela, M., Influence of pH on the toxicity of substituted phenols to fish, *Arch. Environ. Contam. Toxicol.,* 10, 747, 1981.
36. Grohmann, A. and Sobhani, P., Beeinflussung der Giftwirkung von Stoffen auf Fische durch Ionengleichgewiche (Harte des wassers and Toxizitat), *Hydrochem. Hydrogeol.,* 3, 147, 1979.
37. Woodward, D. F., Toxicity of the herbicides dinoseb and picloram to cutthroat (*Salmo clarki*) and lake trout (*Salvelinus namaycush*), *J. Fish. Res. Board Can.,* 33, 1671, 1976.
38. Mauck, W. L., Olson, L. E., and Hogan, J. W., Effects of water quality on deactivation and toxicity of mexacarbate (Zectran) on fish, *Arch. Environ. Contam. Toxicol.,* 6, 385, 1977.
39. Folmar, L. C., Sanders, H. O., and Julin, A. M., Toxicity of the herbicide glyphosate and several of its formulations to fish and aquatic invertebrates, *Arch. Environ. Contam. Toxicol.,* 8, 269, 1979.
40. Marking, L. L. and Bills, T. D., Sensitivity of four species of carp to selected fish toxicants, *N. Am. J. Fish. Manage.,* 1, 51, 1981.
41. Birge, W. J., Black, J. A., Hudson, J. E., and Bruser, D. M., Embryo-larval toxicity tests with organic compounds, in *Aquatic Toxicology,* ASTM STP 667, Marking, L. L. and Kimerle, R. A., Eds., American Society for Testing and Materials, Philadelphia, 1979, 131.
42. Inglis, A. and Davis, E. L., Effects of water hardness on the toxicity of several organic and inorganic herbicides to fish, Fish and Wildlife Service, U.S. Department of the Interior, Washington, D.C., 1972.
43. Finlayson, B. J., Acute toxicities of the herbicides Komeen and hydrothol-191 to golden shiner (*Notemigonus crysoleucas*), *Bull. Environ. Contam. Toxicol.,* 25, 676, 1980.
44. Kawatski, J. A., Ledvina, M. M., and Hansen, C. R., Jr., Acute toxicities of 3-trifluoromethyl-4-nitrophenol (TFM) and 2′,5-dichloro-4′-nitrosalicylanilide (Bayer 73) to larvae of the midge *Chironomus tentans,* in Investigations in Fish Control 57, Fish and Wildlife Service, U.S. Department of the Interior, Washington, D.C., 1975.
45. Marking, L. L. and Olson, L. E., Toxicity of the lampricide 3-trifluoromethyl-4-nitrophenol (TFM) to nontarget fish in static tests, in Investigations in Fish Control 60, Fish and Wildlife Service, U.S. Department of the Interior, Washington, D.C., 1972.
46. Bailey, G. W. and White, J. L., Review of adsorption and desorption of organic pesticides by soil colloids, with implications concerning pesticide bioactivity, *J. Agric. Food Chem.,* 12, 324, 1964.

47. Fredeen, F. J. H., Arnason, A. P., and Berck, B., Adsorption of DDT on suspended solids in river water and its role in black-fly control, *Nature*, 171, 700, 1953.

48. Brungs, W. A. and Bailey, G. W., Influence of suspended solids on the acute toxicity of endrin to fathead minnows, Engineering Extension Series 121, *Eng. Bull. Purdue Univ.*, Proc. 21st Purdue Industrial Waste Conf. Part I., Lafayette, Ind., 1966, 4.

49. Liang, T. T. and Lichtenstein, E. P., Synergism of insecticides by herbicides: effect of environmental factors, *Science*, 186, 1128, 1974.

50. Kok, L. T. and Pathak, M. D., Toxicity of lindane used for asiatic rice borer control to three species of fish, *J. Econ. Entomol.*, 59, 659, 1966.

51. Ferguson, D. E., Ludke, J. L., Wood, J. P., and Prather, J. W., The effects of mud on the bioactivity of pesticides on fishes, *J. Miss. Acad. Sci.*, 11, 219, 1965.

52. Zitko, V., Uptake of chlorinated paraffins and PCB from suspended solids and food by juvenile Atlantic salmon, *Bull. Environ. Contam. Toxicol.*, 12, 406, 1974.

53. Halter, M. T. and Johnson, H. E., A model system to study the desorption and biological availability of PCB in hydrosoils, in *Aquatic Toxicology and Hazard Evaluation,* ASTM STP 634, Mayer, F. L. and Hamelink, J. L., Eds., American Society for Testing and Materials, Philadelphia, 1977, 178.

54. Nau-Ritter, G. M. and Wurster, C. F., Sorption of polychlorinated biphenyls (PCB) to clay particulates and effects of desorption on phytoplankton, *Water Res.*, 17, 383, 1983.

55. Nau-Ritter, G. M., Wurster, C. F., and Rowland, R. G., Polychlorinated biphenyls (PCB) desorbed from clay particles inhibit photosynthesis by natural phytoplankton communities, *Environ. Pollut.*, 28, 177, 1982.

56. Bopp, R. F., Simpson, H. J., Olson, C. R., and Kostyk, N., Polychlorinated biphenyls in sediments of the tidal Hudson River, New York, *Environ. Sci. Technol.*, 15, 210, 1981.

57. Vanderford, M. J. and Hamelink, J. L., Influence of environmental factors on pesticide levels in sport fish, *Pestic. Monit. J.*, 11, 138, 1977.

58. Herzel, F. and Murty, A. S., Do carrier solvents enhance the water solubility of hydrophobic compounds?, *Bull. Environ. Contam. Toxicol.*, 32, 53, 1984.

59. Brown, V. M., Shurben, D. G., and Fawell, J. K., The acute toxicity of phenol to rainbow trout in saline waters, *Water Res.*, 1, 683, 1967.

60. Murphy, P. G., Effects of salinity on uptake of DDT, DDE and DDD by fish, *Bull. Environ. Contam. Toxicol.*, 5, 404, 1970.

61. Tulp, M. Th. M., Hay, K., Carson, W. G., Zitko, V., and Hutzinger, O., Effect of salinity on uptake of ^{14}C-2,2',4,5,5'-pentachlorobiphenyl by juvenile Atlantic salmon, *Chemosphere*, 8, 243, 1979.

62. Pickering, Q. H., Some effects of dissolved oxygen concentrations upon the toxicity of zinc to the bluegill, *Lepomis macrochirus, Water Res.*, 2, 187, 1968.

63. Adelman, I. R. and Smith, L. L. Jr., Toxicity of hydrogen sulfide to goldfish (*Carassius auratus*) as influenced by temperature, oxygen and bioassay techniques, *J. Fish. Res. Board Can.*, 29, 1309, 1972.

64. Odum, E. P. and Sumerford, W. T., Comparative toxicity of DDT and four analogues to goldfish, *Gambusia* and *Cluex* larvae, *Science*, 104, 480, 1946.

65. Wershaw, R. L., Burcar, P. J., and Goldberg, M. C., Interaction of pesticides with natural organic material, *Environ. Sci. Technol.*, 3, 271, 1969.

66. Hamelink, J. L. and Waybrant, R. C., DDE and lindane in a large-scale model lentic ecosystem, *Trans. Am. Fish. Soc.*, 105, 124, 1976.

67. Mount, D. I., Adequacy of laboratory data for protecting aquatic communities, in *Analyzing and Hazard Evaluation Process,* Dickson, K. L., Maki, A. W., and Cairns, J., Jr., Eds., American Fisheries Society, Washington, D.C., 1979, 112.

68. Ambient Water Quality Criteria for DDT, EPA 440/5-80-038, Water Office of Regulations and Standards Criteria and Standards Division, Environmental Protection Agency, Washington, D.C., 1980.

69. Gills, S. S., In vitro metabolism of carbofuran by liver microsomes of the paddy field fish *Trichogaster pectoralis, Bull. Environ. Contam. Toxicol.*, 25, 697, 1980.

70. Muir, D. C. G., Grift, N. P., Blouw, A. P., and Lockhart, W. L., Environmental dynamics of phosphate esters. I. Uptake and bioaccumulation of triphenyl phosphate by rainbow trout, *Chemosphere*, 9, 525, 1980.

71. Muir, D. C. G. and Grift, N. P., Environmental dynamics of phosphate esters. II. Uptake and bioaccumulation of 2-ethylhexyl diphenyl phosphate and diphenyl phosphate by fish, *Chemosphere*, 10, 847, 1981.

72. Muir, D. C. G., Grift, N. P., Townsed, B. E., Metner, D. A., and Lockhart, W. L., Comparison of the uptake and bioconcentration of fluridone and terbutryn by rainbow trout and *Chironomus tentans* in sediment and water systems, *Arch. Environ. Contam. Toxicol.*, 11, 595, 1982.

73. Lockhart, W. L., Metner, D. A., Blouw, A. P., and Muir, D. C. G., Prediction of biological availability of organic chemical pollutants to aquatic animals and plants, in *Aquatic Toxicology and Hazard Assessment: Fifth Conference,* ASTM STP 766, Pearson, J. G., Foster, R. B., and Bishop, W. E., Eds., American Society for Testing and Materials, Philadelphia, 1982, 259.

74. Fox, P. J. and Matthiessen, P., Acute toxicity to fish to low-dose aerosol applications of endosulfan to control tsetse fly in the Okavango Delta, Botswana, *Environ. Pollut.,* 27, 129, 1982.

75. Seelye, J. G. and Mac, M. J., Size-specific mortality in fry of lake trout (*Salvelinus namaycush*) from Lake Michigan, *Bull. Environ. Contam. Toxicol.,* 27, 376, 1981.

76. Seelye, J. G., Hesselberg, R. J., and Mac, M. J., Accumulation by fish of contaminants released from dredged sediments, *Environ. Sci. Technol.,* 16, 459, 1982.

77. Muir, D. C. G. and Grift, N. P., Fate of fluridone in sediment and water in laboratory and field experiments, *J. Agric. Food Chem.,* 30, 238, 1982.

78. Bill, T. D., Marking, L. L., and Olson, L. E., Effects of residues of the polychlorinated biphenyl Aroclor 1254 on the sensitivity of rainbow trout to selected environmental contaminants, *Prog. Fish-Culturist,* 39, 150, 1977.

79. Chandler, J. H., Jr. and Marking, L. L., Toxicity of rotenone to selected aquatic invertebrates and frog larvae, *Prog. Fish-Culturist,* 44, 78, 1982.

80. Hamelink, J., Bioavailability of chemicals in aquatic environments, in *Biotransformation and Fate of Chemicals in the Environment,* Maki, A. W., Dickson, K. L., and Cairns, J., Jr., Eds., American Society of Microbiology, Washington, D.C., 1980, 56.

Chapter 3

TOXICITY OF FORMULATIONS, ISOMERS, AND DEGRADATION PRODUCTS

I. INTRODUCTION

Frequently, compounds of highest purity like analytical-grade reagents or technical-grade compounds are employed to test the toxicity of pesticides to the aquatic organisms. It must, however, be remembered that pure compounds are seldom, if ever, used in nature. The pesticides are formulated as emulsifiable concentrates (EC), wettable powders (WP), granulated preparations, dusts, aerosols, and the like, which help in dispersing the compound better on the target sites. Various types of inert materials are used as the carriers, emulsifiers, diluents, etc. to disperse the active ingredient (a.i.) on the target sites. These inert materials remain rarely inert, and are often in themselves toxic, or may interact with the a.i., the resultant toxicity being additive, more-than-additive, or less-than-additive (see Chapter 4).

The formulated pesticide used in nature is never transported *in toto* to the aquatic environment. After the application of the formulation, it is the physical and chemical properties of the pesticide molecule, outlined in Chapter 1, Volume I, that influence the transport of pesticides to the hydrosphere. Nevertheless, because of spray drift during application and also direct application of the pesticide formulations to the water bodies (for purposes of weed and nuisance insect control), the aquatic organisms may be occasionally exposed directly to the formulations and the inert ingredients therein. Hence, it is imperative to test the toxicity of the formulations as they are used in nature, along with testing the toxicity of the pure compound, to the nontarget organisms in the aquatic ecosystem.

It is also necessary to test the toxicity of the degradation products, especially that of the primary ones in nature, which sometimes may be as toxic as, if not more toxic than, the parent compounds. Yet another aspect that needs to be evaluated is the differential toxicity of the isomers, when the technical-grade material is composed of two or more isomers as in the case of DDT, endosulfan, HCH, etc. The toxicity of the formulations, degradation products, and isomers is reviewed hereunder.

II. TOXICITY OF FORMULATIONS

Various types of pesticide formulations have been excellently reviewed by Melnikov.[1] In 1971, it was estimated that there were about 100,000 formulations of 900 pesticides in use in different countries.

Technical lindane was more toxic than a dust formulation, the 96-h LC 50 of technical-grade material and the dust formulation to the bluegill being 57 and 138 $\mu g/\ell$.[2] The 96-h LC 50 of technical DDT and 25% EC to bluegills was, respectively, 3.4 and 9 $\mu g/\ell$, and to *Daphnia* 1.1 and 1.7 $\mu g/\ell$.[2] The 48-h LC 50s of a commercial mixture (75% *p,p'*-DDT and 20% *o,p'*-DDT) of DDT and WP were 0.5 to 2.5 and 8 $\mu g/\ell$, respectively. The 24-h LC 50 of a paste formulation and that of a liquid formulation with emulsifiers was 10.7, and 20 to 110 $\mu g/\ell$, respectively.[3] A similar reduction in the toxicity of EC formulations was reported in the case of endrin and chlordane. The 96-h LC 50s of technical-grade material and EC formulations (19.5% a.i.) of the former to fathead minnows were 1.1 and 2.9 $\mu g/\ell$, respectively, and to bluegills, 0.66 and 3.7 $\mu g/\ell$. The 96-h LC 50s of 100% pure chlordane and 75% EC to fathead minnows were 52 and 170 $\mu g/\ell$.[4] Compared with the toxicity of technical-grade endrin (cited above), the toxicity of WP formulation to fatheads was low (96-h LC 50 — 2.6 $\mu g/\ell$). No difference was noticed in the toxicity of technical-grade heptachlor (consisting of 65% heptachlor, 22% trans-chlordane, 2% cis-chlordane, and 2% nonachlor) and analytical-grade heptachlor (99% pure) to *Leiostomus xanthurus*, an estuarine fish.[5] On the other hand, an EC formulation of methoxychlor was more toxic than a particulate formulation to fingerling rainbow trout.[6]

Studies in our laboratory on the various types of formulations of endosulfan showed that 35% EC formulation was more toxic than the technical-grade material to fish, whereas 4% dust formulation was less toxic than the technical material. The 96-h LC 50s of technical-grade endosulfan to *Labeo rohita* and *Channa punctata* were 1.1 and 4.8 $\mu g/\ell$, those of 35% — EC 1 and 2.5 $\mu g/\ell$, and those of 4% dust formulation — 1.3 and 16 $\mu g/\ell$.[7,8]

A decrease in toxicity and consequent doubling of the 96-h LC 50 value by the addition of 3.5% inert ingredient to technical parathion was reported.[9] Decreased toxicity of an EC formulation of parathion to fathead minnows, bluegills, and goldfish in comparison with that of the technical parathion was reported by Henderson and Pickering.[9] Similar results of decreased toxicity in the case of 25% EC formulation of parathion in comparison with technical-grade (99%) parathion, to fathead minnows, bluegills, and goldfish was noted by Pickering et al.[10] A sixfold decrease in the toxicity of 20% malathion to bluegills, a fourfold decrease in the toxicity of EPN 25% WP to fathead minnows and bluegills, and a slight but significant reduction in the toxicity of Delnav® 47% EC in comparison with their respective technical-grade materials was also noted in the same study. A 14-fold increase in the toxicity of a commercial formulation of malathion, over that of the technical-grade, was reported by Sailatha et al.[11]

Sesamex, present in many of the organophosphate (OP) compound formulations at about 1% concentration, normally increases the toxicity of the a.i. (this action was formerly called synergism — see Chapter 4), but in the case of parathion, sesamex decreased its toxicity to mosquitofish.[12]

Aminocarb EC formulation — metacil — is more toxic than technical aminocarb to the Atlantic salmon, the 96-h LC 50 being 3.5 and 8.7 mg/ℓ, respectively. Increased toxicity of the formulation was attributable to nonylphenol, a component of the formulation. Another component of the formulation, 585 oil, had no effect.[13] Likewise, formulated products were more toxic than the parent compounds in the case of synthetic pyrethroids; the extent of the higher toxicity of the formulated product over that of the technical-grade material ranged from 2 in the case of permethrin to 9 in the case of fenpropanate.[14] Also, with technical-grade pyrethroids, mortality occurred quickly or not at all, whereas with formulations, mortality was more evenly distributed throughout the 24-hr study period.

While studying the behavioral changes induced by Dimilin-G1®, an insect growth regulator, it was observed that the changes produced by the carrier, Florex®, in the male Atlantic salmon parr, were the same as those produced by the pesticide.[15] Although Sanders and Walsh[16] stated that purified 3-trifluoromethyl-4-nitrophenol (TFM) (97.5%) was twice as toxic as field grade TFM (35.7%) to crayfish, when the a.i. present is taken into consideration, the reverse appears to be true, both at the end of 24 and 96 hr. Marking and Olson[17] found field grade TFM more toxic than the pure compound.

The toxicity of ten amine 2,4-dichlorophenoxyacetic acid (2,4-D) formulations to the bluegill sunfish was reported by Hughes and Davis.[18] The 48-h LC 50 of different amines ranged from 0.8 mg/ℓ (isopropyl ester) to 840 mg/ℓ of alkonalamine. Apparent differences in the same type of formulation produced by the same processing plant, but belonging to different batches, was noted. The 48-h LC 50 value of five different dimethylamine preparations from five different formulating factories ranged from 166 to 458 mg/ℓ. The ester formulations of 2,4-D (ethyl, isopropyl, butyl, etc.) proved to be highly toxic in comparison with other formulations, the 48-h LC 50 ranging from 0.8 mg/ℓ of isopropyl ether, to 8.8 to 59.7 mg/ℓ of isooctyl ether. In another study on the toxicity of various formulations of 2,4-D to the salmonids, butyl ester was found to be the most toxic of the ester formulations, and the isooctyl ester the least toxic;[19] propylene glycol butyl ester (PGBE) and isooctyl ester were the least toxic.[19] Part of the reason for the difference in the toxicity of the various esters probably lies in the ester structure itself. Zepp et al.[20] reported that esters of 2,4-D that possess ether linkages near the carboxyl group hydrolyze more easily than those with hydrocarbon chain esters.

In the case of 2,4,5-T (2,4,5-trichlorophenoxyacetic acid), too, dimethylamine and isooctyl ester formulations proved least toxic, whereas butoxyethanol ester was more toxic. In the case of 2-(2,4-DP) and Silvex® also, butoxyethanol ether formulation was the most toxic among the formulations tested. Isooctyl ether, which has intermediate toxicity among all the formulations of 2,4-D, 2,4,5-T and 2-(2,4-DP) was toxic in the case of Silvex®.[18]

Folmar et al.[21] studied the relative toxicity of technical-grade glyphosate, isopropylamine salt of glyphosate, a field formulation with a surfactant, and another formulation called Roundup® surfactant to four species of aquatic invertebrates, and four species of fish. Technical-grade glyphosate was considerably less toxic than the two formulations which had similar toxicities. The surfactant was the primary toxic agent of the formulations. Granular formulations of some herbicides were less toxic to fish fry. Similarly, water-soluble derivatives of Silvex®, 2,4,5-T, and 2,4-D were less toxic than ester derivatives.[22]

As the foregoing account reveals, the formulations rarely have the same toxicity as the pure compounds. When the inert ingredients themselves have no toxicity, the toxicity of a formulation depends on the amount of the active ingredient present, as has been reported for the formulations of simazine;[3] the higher the content of the a.i., the higher the toxicity of the formulation. When the additives for formulation are not inert, the toxicity of formulations differs widely, depending on the toxicity of the additives.

Controlled release pesticides (CRP), also called encapsulated pesticides, are being increasingly used to ensure slow, continuous, and effective release of the pesticide into the environment, and long-lasting control of the target species. Jarvinen and Tanner[23] studied the toxicity of technical-grade methyl parathion, Dursban®, and diazinon and their CRPs, viz., Penncap-M®, Dursban® 10 CR, and Knox out® 2 FM, respectively, to fathead minnows under static and flow-through conditions. The toxicity of the aged solutions of these compounds in water also was studied. With increase in the age of solutions under static conditions, toxicity of methyl parathion, diazinon, and Penncap-M® increased. The toxicity of technical-grade Dursban® and Knox out® 2 FM remained constant with time, whereas that of Dursban® 10 CR decreased with time. The increase in the toxicity of aged products observed in some cases was due to the accumulation of breakdown products that are more toxic than the parent compounds. Only traces of methyl paraoxan was present in the case of both technical-grade and formulated compound, the increased toxicity being attributable to the accumulation of p-nitrophenol. The oxygen analog of diazinon — diazoxon — which is known to be more toxic than the parent compound, accounted for the increased toxicity of the aged solutions of technical-grade diazinon. The encapsulated formulation of methyl parathion was 45 to 60% less toxic than technical-grade under static conditions, whereas it was only 22% less toxic in flow-through conditions because of lower build-up of toxic degradation products under flow-through conditions. The 96-h LC 50 under static conditions for technical-grade products of methyl parathion, Dursban®, and diazinon was (mg/ℓ) 4.46, 0.17, and 4.3, respectively, whereas that of encapsulated formulations was 8.17, 0.13, and 4.3, respectively. Under flow-through conditions the 96-h LC 50 of methyl parathion, Dursban®, and diazinon was 5.36, 0.14, and 6.9; whereas that of formulations of methyl parathion, Dursban®, and diazinon was 6.91 and 0.12. For diazinon-encapsulated product no value could be obtained, as not enough pesticide was released from the formulation under flow-through conditions to cause mortality of fish. In general, long-term toxicity of both technical-grade compounds and CRPs was similar.

Another problem that was not encountered with the technical-grade material, but only with the formulation, arose during experiments to devise means of disposing of small quantities of pesticide formulations and decontaminating empty containers. At pH 8 to 10, pure diazinon could be completely degraded by treatment with sodium hypochlorite. Diazinon EC retained most of its toxicity, even after treatment with sodium hypochlorite. The toxic components of the EC formulation were identified as a heavy, aromatic naphtha constituent and also chlorinated, aromatic hydrocarbons produced in the reaction mixture. These components were as toxic as the EC formulation. Thus, in any attempt at disposal of pesticides (and the substances that are to be disposed of are always formulations), it is not enough if the compound is detoxified. It is equally necessary to ensure that the formulations are also made nontoxic to the nontarget organisms.[24]

III. TOXICITY OF ISOMERS

Between the two isomers of DDT, p,p'=DDT is more toxic to both the target and nontarget organisms; p,p'-DDT was six times as toxic as the other isomer to fish.[3] Pure p,p'-DDT was about twice as toxic as technical-grade DDT to fathead minnows.[4] The presence of o,p'-DDT in the technical-grade material (70 to 75% p,p'-DDT and about 20% o,p'-DDT) only seems to lessen the toxicity of technical material in comparison with that of p,p'-DDT. When given through diet, p,p'-DDT was more toxic than the technical-grade DDT to chinook and coho salmon. Salmon which were given 100 mg p,p'-DDT/kg diet accumulated significantly greater tissue concentrations of DDT and its metabolites than fish that were given 100 mg technical-grade DDT/kg.[25]

Technical lindane was less toxic than an equivalent concentration of γ-HCH (hexachlorocyclohexane) present along with other isomers in technical-grade HCH. The other isomers of HCH apparently lessened the toxicity of γ-HCH in the technical-grade material. While 0.1 mg/ℓ lindane killed all the fathead minnows in a tank in 24 hr, 3.2 mg/ℓ technical HCH and 0.1 mg/ℓ lindane together in the same tank caused no fish mortality in 96 hr.[4] Similar observations were made by Schimmel et al.[26] while testing the toxicity of lindane and technical-grade HCH to estuarine animals. The 96-h LC 50 of lindane and technical HCH to the pinfish was 30.6 and 86.4 μg/ℓ, respectively; similarly, their toxicity to the crustacean, *Peanaeus duorarum* was, respectively, 0.17 and 0.34 μg/ℓ.[26]

In our laboratory we investigated the toxicity of isomers-a and -b of endosulfan, to three species of freshwater fish. Isomer-a was consistently more toxic than the technical-grade endosulfan and isomer-b was always less toxic than the technical-grade endosulfan.[7,8,27] The 96-h LC 50s of isomers-a and -b were, respectively, 0.33 and 7.1 μg/ℓ to *Labeo rohita*, 0.16 and 6.6 μg/ℓ to *Channa punctata*, and 0.6 and 8.8 μg/ℓ to *Cirrhinus mrigala*. In a study on the biochemical changes induced by both the isomers, the changes induced by isomer-a were more striking and of a greater magnitude than those induced by isomer-b.[28]

Among the synthetic pyrethroids, (1R)-*cis* isomers were more toxic than (1R)-*trans* isomers.[29] S-Methyl fenitrothion, an impurity present in technical fenitrothion, was as toxic as fenitrothion to juvenile Atlantic salmon.[30]

The greater toxicity of one of the isomers present in the technical-grade material and formulations to fish, if coupled with a greater mobility of that isomer in the soil (and hence consequent greater transportation to the aquatic environment) as has been reported for isomer-a of endosulfan, may prove to be a greater hazard to the aquatic life in areas where such a transport is likely to occur.

IV. TOXICITY OF DEGRADATION PRODUCTS

The natural degradation products of pesticides, in general, are less toxic than the parent compounds. But occasionally, the primary degradation product may be more toxic. The greater toxicity of photoisomers of some of the cylodienes than their normal isomers has been mentioned in Chapter 1. Likewise, the hydrolysis product of malathion — diethyl fumarate — was more toxic than the parent compound to the fathead minnows.[31] The breakdown product of mexacarbate — 4-amino-3,5-xylelon — was significantly more toxic than the parent compound to several species of fish.[32]

Carbaryl has found very wide use, especially after the ban on DDT in the western countries. The primary breakdown product of carbaryl (1-naphthol) is believed to be nontoxic in the terrestrial environment. But in the few cases where its toxicity to the aquatic organisms has been tested, it was found to be more toxic than the parent compound. Stewart et al.[33] were the first to report the greater toxicity of 1-naphthol to molluscs and three species of marine fish. Subsequently, Butler et al.[34] reported that the growth of juvenile cockleclams (*Clinocardium nuttalli)* was reduced more by 1-naphthol than by carbaryl, and also that 1-naphthol was more toxic to juvenile clams. In our laboratory, we found that 1-naphthol was always more toxic than the parent compound to the freshwater fish we tested.[35-37] For six species of fish, 1-naphthol was about 2 times as toxic as carbaryl, but in the case of *Mystus cavasius,* we found the degradation product 14 times as toxic as the parent compound (96-h LC 50s being, respectively, 0.33 and 4.6 mg/ℓ).

The above discussion clearly reveals that studies on the toxicity of isomers and degradation products are very few. Particularly with the latter, there is a greater need for the evaluation of their toxicity to the aquatic organisms, because in choosing an environmentally less hazardous chemical, the toxicity of the main chemical alone is assessed. If there are any studies at all on the toxicity of the primary degradation products, often they are confined to the target and nontarget organisms in the terrestrial environment. Carbaryl is a case in point. Since the evaluation of the toxicity of the degradation products in the terrestrial environment does not reveal their toxicity to the aquatic organisms, it is imperative that aquatic toxicologists pay more attention to this aspect, too.

V. CONCLUSIONS

While evaluating the toxicity of pesticides, it is not enough if the toxicity of the pure compound is assessed. The aquatic organisms may be occasionally exposed to the formulations. In general, emulsifiable concentrates are more toxic than the technical-grade material, whereas dust and powder formulations are less toxic. Also, different isomers of a compound may be mobile in the soils to different degrees; thereby aquatic organisms may be exposed not to the pure compound, but only to one or more of the isomers to a greater extent. Further, while selecting a pesticide for use, the toxicity of the degradation product is evaluated more often than not, in the terrestrial environment. Such degradation products may turn out to be more toxic than the parent compounds to the aquatic organisms, as in the case of 1-naphthol. Hence, aquatic toxicologists should study the toxicity of primary degradation products in the aquatic environment, too.

REFERENCES

1. Melnikov, N. N., Chemistry of pesticides, *Residue Rev.*, 36, 480, 1971.
2. Randall, W. F., Dennis, W. H., and Warner, M. C., Acute toxicity of dechlorinated DDT, chlordane and lindane to bluegills (*Lepomis macrochirus*) and *Daphnia magna, Bull. Environ. Contam. Toxicol.*, 21, 849, 1979.
3. Alabaster, J. S., Survival of fish in 164 herbicides, insecticides, fungicides, wetting agents and miscellaneous substances, *Int. Pest Control*, 2, 29, 1969.
4. Henderson, C., Pickering, Q. H., and Tarzwell, C. M., Relative toxicity of ten chlorinated hydrocarbon insecticides to four species of fish, *Trans. Am. Fish. Soc.*, 88, 23, 1959.
5. Schimmel, S. C., Patrick, J. M., Jr., and Forester, J., Heptachlor: toxicity to and uptake by several estuarine organisms, *J. Toxicol. Environ. Health*, 1, 955, 1976.
6. Sebastien, R. J. and Lockhart, W. L., The influence of formulation on toxicity and availability of a pesticide (methoxychlor) to black fly larvae (Diptera: Simuliidae), some non-target aquatic insects and fish, *Can. Entomol.*, 113, 281, 1981.
7. Rao, D. M. R., Priyamvada Devi, A., and Murty, A. S., Relative toxicity of endosulfan, its isomers, and formulated products to the freshwater fish, *Labeo rohita, J. Toxicol. Environ. Health*, 6, 825, 1980.
8. Priyamvada Devi, A., Rao, D. M. R., Tilak, K. S., and Murty, A. S., Relative toxicity of the technical grade material, isomers, and formulations of endosulfan to the fish *Channa punctata, Bull. Environ. Contam. Toxicol.*, 27, 239, 1981.
9. Henderson, C. and Pickering, Q. H., Toxicity of organicphosphorus insecticides to fish, *Trans. Am. Fish. Soc.*, 87, 39, 1957.
10. Pickering, Q. H., Henderson, C., and Lemke, A. E., The toxicity of organicphosphorus insecticides to different species of warmwater fishes, *Trans. Am. Fish. Soc.*, 91, 175, 1962.
11. Sailatha, D., Kabeer Ahammad Sahib, I., and Ramana Rao, K. V., Toxicity of technical and commercial grade malathion to the fish, *Tilapia mossambica* (Peters), *Proc. Indian Acad. Sci.*, 90, 87, 1981.
12. Gibson, J. R. and Ludke, J. L., Effect of sesamex on brain acetylcholinesterase inhibition by parathion in fishes, *Bull. Environ. Contam. Toxicol.*, 6, 97, 1971.
13. McLeese, D. W., Zitko, V., Metcalfe, C. D., and Sergeant, D. B., Lethality of aminocarb and the components of the aminocarb formulation of juvenile Atlantic salmon, marine invertebrates and a freshwater clam, *Chemosphere*, 9, 79, 1980.
14. Coats, J. R. and O'Donnell-Jefferey, N. L., Toxicity of four synthetic pyrethroid insecticides to rainbow trout, *Bull. Environ. Contam. Toxicol.*, 23, 250, 1979.
15. Granett, J., Morang, S., and Hatch, R., Reduced movement of precocious male Atlantic salmon parr into sublethal Dimilin-G₁, and carrier concentrations, *Bull. Environ. Contam. Toxicol.*, 19, 462, 1978.
16. Sanders, H. O. and Walsh, D. F., Toxicity and residue dynamics of the lampricide 3-trifluoromethyl-4-nitrophenol (TFM) in aquatic invertebrates, in Investigations in Fish Control 59, Fish and Wildlife Service, U.S. Department of the Interior, Washington, D.C., 1975.
17. Marking, L. L. and Olson, L. E., Toxicity of the lampricide 3-trifluoromethyl-4-nitrophenol (TFM) to nontarget fish in static tests, in Investigations in Fish Control 60, Fish and Wildlife Service, U.S. Department of the Interior, Washington, D.C., 1975.
18. Hughes, J. S. and Davis, J. T., Variations in toxicity to bluegill sunfish of phenoxy herbicides, *Weeds*, 11, 50, 1963.
19. Meehan, W. R., Norris, L. A., and Sears, H. S., Toxicity of various formulations of 2,4-D to salmonids in southeast Alaska, *J. Fish. Res. Board Can.*, 31, 480, 1974.
20. Zepp, R. G., Wolfe, N. L., Gordon, J. A., and Baughman, G. L., Dynamics of 2,4-D esters in surface waters hydrolysis, photolysis, and vaporization, *Environ. Sci. Technol.*, 9, 1144, 1975.

21. Folmar, L. C., Sanders, H. O., and Julin, A. M., Toxicity of the herbicide glyphosate and several of its formulations to fish and aquatic invertebrates, *Arch. Environ. Contam. Toxicol.*, 8, 269, 1979.

22. Hiltibran, R. C., Effects of some herbicides on fertilized fish eggs and fry, *Trans. Am. Fish. Soc.*, 96, 414, 1967.

23. Jarvinen, A. W. and Tanner, D. K., Toxicity of selected controlled release and corresponding unformulated technical grade pesticides to the fathead minnow *Pimephales promelas*, *Environ. Pollut.*, 27, 179, 1982.

24. Dennis, W. H., Jr., Meier, E. P., Randall, W. F., Rosencrance, A. B., and Rosenblatt, D. H., Degradation of diazinon by sodium hypochlorite. Chemistry and aquatic toxicity, *Environ. Sci. Technol.*, 13, 594, 1979.

25. Buhler, D. R., Rasmusson, M. E., and Shanks, W. E., Chronic oral DDT toxicity in juvenile coho and chinook salmon, *Toxicol. Appl. Pharmacol.*, 14, 535, 1969.

26. Schimmel, S. C., Patrick, J. M., Jr., and Forester, J., Toxicity and bioconcentration of BHC and lindane in selected estuarine animals, *Arch. Environ. Contam. Toxicol.*, 6, 355, 1977.

27. Fromm, P. O., Studies on renal and extra-renal excretion in a freshwater teleost, *Salmo gairdneri*, *Comp. Biochem. Physiol.*, 10, 121, 1963.

28. Murty, A. S. and Priyamvada Devi, A., The effect of endosulfan and its isomers on tissue protein, glycogen, and lipids in the fish *Channa punctata*, *Pestic. Biochem. Physiol.*, 17, 280, 1982.

29. Zitko, V., McLeese, D. W., Metcalfe, C. D., and Carson, W. G., Toxicity of permethrin, decamethrin, and related pyrethroids to salmon lobster, *Bull. Environ. Contam. Toxicol.*, 21, 338, 1979.

30. Zitko, V. and Cunningham, T. D., Fish toxicity of S-methyl fenitrothion, *Bull. Environ. Contam. Toxicol.*, 14, 19, 1975.

31. Bender, M. E., The toxicity of the hydrolysis and breakdown products of malathion to the fathead minnow (*Pimephales promelas*, Rafinesque), *Water Res.*, 3, 571, 1969.

32. Mauck, W. L., Olson, L. E., and Hogan, J. W., Effects of water quality on deactivation and toxicity of mexacarbate (Zectran) to fish, *Arch. Environ. Contam. Toxicol.*, 6, 385, 1977.

33. Stewart, N. E., Millemann, R. E., and Breese, W. P., Acute toxicity of the insecticide Sevin and its hydrolytic product 1-naphthol to some marine organisms, *Trans. Am. Fish. Soc.*, 96, 25, 1967.

34. Butler, J. A., Millemann, R. E., and Stewart, N. E., Effects of the insecticide Sevin on survival and growth of the cockle clam *Clinocardium nuttalli*, *J. Fish. Res. Board Can.*, 25, 1621, 1968.

35. Tilak, K. S., Mohana Ranga Rao, D., Priyamvada Devi, A., and Murty, A. S., Toxicity of carbaryl and 1-naphthol to the freshwater fish *Labeo rohita*, *Ind. J. Exp. Biol.*, 18, 75, 1980.

36. Tilak, K. S., Mohana Ranga Rao, D., Priyamvada Devi, A., and Murty, A. S., Toxicity of carbaryl and 1-naphthol to four species of freshwater fish, *J. Biosci.*, 3, 457, 1981.

37. Mohana Rao, D., Murty, A. S., and Ananda Swarup, P., Relative toxicity of technical grade and formulated carbaryl and 1-naphthol to, and carbaryl-induced biochemical changes in the fish, *Cirrhinus mrigala*, *Environ. Pollut.*, 34, 47, 1984.

Chapter 4

JOINT ACTION OF PESTICIDE MIXTURES

I. INTRODUCTION

In nature, organisms are exposed not to a single toxicant, but often to more than one chemical present either in considerable amounts or in traces. Experimentally, also, the desire to enhance the effects of individual compounds by applying two or more of them together calls for the application of pesticide mixtures. The identification of the contribution of each of these chemicals to the overall toxic effect and the interaction between any two or a group of toxicants has been much discussed. For a long time it has been realized that interactions between chemicals may either increase or decrease the overall effect, i.e., the resulting action is more or less than the simple combined effect, or there may not be any interaction at all between the two. Further, the situation is complicated when one of the components is not toxic by itself, but in its presence the other component acts as though more of it is acting than is actually present.

Various terms like "synergism", "potentiation", "antagonism", etc. have been in use to describe the various situations referred to above. These terms have not succeeded to define the diverse processes and clarify the picture for the following reasons: (1) different workers have used different terms to denote the same thing; (2) the same term has been used by different workers to describe different situations; (3) although a term may be apparently suitable to describe a particular situation, the explanation does not lend itself to the formation of a suitable "null hypothesis" and its testing, i.e., the explanation is not amenable to statistical analysis; (4) a particular definition may be apparently suitable, but the model fails when one attempts to explain the resultant action in terms of the competition at the site or sites of action (target organs or sites); and (5) depending on the way they are defined, the processes like synergism and antagonism could occur simultaneously, as has been pointed out by Morse.[1] Thus, when one attempts to understand or explain the joint action of pesticides, the picture is confusing and apparently good terms have fallen into disrepute.

Before examining the laboratory and field experiments designed to test the joint action of pesticides on fish, a few reviews that have greatly helped in clearing the confusion that exists in this subject are summarized.

II. THEORY

Sprague[2] considered the scheme of Gaddum[3] as precise, without any contradictions, and suitable to fit toxicity experiments in the aquatic medium. In this scheme, the joint action is defined as additive, less-than-additive, more-than-additive, or antagonistic. If half of the concentration of toxicant A needed to produce a given response and half of the concentration of toxicant B needed to produce the same response together produce that response, the action of toxicants A and B is additive. If this combination causes more than that response, the resultant action is more-than-additive, and if it causes less than that response, then it is less-than-additive or antagonistic, or without interaction. Illustrating his point, Sprague explained that supposing that one toxic unit of toxicant A or B is needed to produce a particular response, then if the same response is produced by combinations of toxicant A and B at concentrations that meet within the square in Figure 1, the toxicants are helping each other and the resultant action is called "joint action". He further explained that three types of joint action can be recognized, viz., if the response is produced by a combination of concentrations that meet at points on the indicated diagonal within the square, it is said to be additive. If

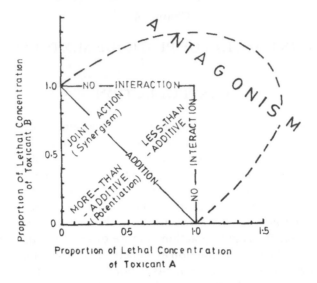

FIGURE 1.   Diagram showing the terms used to describe the combined effects of two toxicants. For details, see text. (From Sprague, J. B., *Water Res.*, 4, 3, 1970. Copyright 1970, Pergamon Press, Ltd. With permission.)

the response is produced by a combination of concentrations that meet each other within the lower triangle, e.g., 0.5A + 0.2B, the effect is more-than-additive. If they meet in the upper right triangle, the action is still joint, but less-than-additive. If one toxic unit concentration of A or B is necessary to produce the response, irrespective of the concentration of the other, then there is no interaction. If more than one toxic unit concentration of either of the toxicants is needed, because of the presence of the other toxicant, then the effect is antagonistic.

More than his attempt at explaining the concept of joint action, Sprague's most important contribution is his clear emphasis that the two terms, synergism and potentiation, should be avoided since they have been defined in more than one way by different workers. He also pointed out certain incongruities in other types of interpretation of joint action. Despite Sprague's clear exposition, it is unfortunate that some authors have persisted in using terms like synergism even today.[4] Akobundu et al.[5] also emphasized the confusion that prevails about terms like synergism. According to them, entomologists define the term as "a substantially more than additive toxic action of two substances used together." This term is applied to situations where one of the components of a mixture is inactive at the concentration employed, but the mixture is more potent than when the other (active) component is used alone. The inactive component in this case is termed the "synergist". It was also pointed out that the definition of synergism as a "co-operative action of different chemicals such that the total effect is greater than the sum of the independent effects" has failed to receive universal acceptance because it cannot be readily evaluated by experimentation.[5] Even as they stressed the untenability of some of these definitions and concepts, Akobundu et al. used the term "synergism" when both the components are active and the combined effect is more than the sum of the two individual effects. They used the term "enhancement" to describe the situation which is usually described as synergism by entomologists, i.e., one of the components is inactive but somehow increases the effect of the other. Schubert et al.[4] proposed a rapid systematic procedure for testing the combined action of two or more toxicants. They also used the terms synergism, antagonism, etc.

with all their attendant deficiencies pointed out previously. Their model was subsequently followed by others, too.

The confusion about what may be considered as synergistic action was also pointed out by Morse[1] who stated, "Despite the long history of the study of mixtures and progress throughout this century in their systematic analytical evaluation, unambiguous definition of these terms has remained elusive." She also pointed out that to test the occurrence of synergism, a null hypothesis has to be postulated, which is not at all difficult, if only one of the components of a mixture affects the organism. If, on the other hand, more than one of the components is separately active, then the null hypothesis is harder to define. She stressed that if synergism is meant to imply that the presence of one component increases the effect of the other, then both synergism and antagonism can occur simultaneously. Morse also pointed out that the area in the rectangle (Figure 1) termed "less-than-additive" by Sprague[2] is also termed "synergism" or "antagonism" by others. Morse[1] concluded, "The limited usefulness of the terms synergism and antagonism, and the prevalence of contradictory definitions support the recommendations . . . to avoid them altogether, seeking instead 'unambiguous mathematical statements to describe the possible results of combining chemical agents.'"[1]

The lack of agreement over the meaning of the various terms is attributable to the failure to define or recognize appropriate models to represent the absence of synergism or antagonism. Discussing the different models proposed by others for examining the joint action, Morse stated that for many years the models on joint action of mixtures were in terms of the doses administered and net effects directly observed. More recently, models have been proposed which take into account the underlying components of joint action, including the effect of one substance on another with respect to the site of action, elimination, metabolism, competition for receptor sites, speed of action, and interaction and so on. She discussed in detail the shortcomings of the various models proposed, including that of Akobundu et al.[5]

Morse[1] discussed two reference models, viz., the additive dose model (ADM) and the multiplicative survival model (MSM). In the former, at any particular level of response, the relative potency of the components, when acting alone, establishes scales of equivalent doses in which if one component is replaced wholly or partly by the other, the predicted response is unchanged. In ADM, the contours of equal response (isoboles) are straight lines. The ADM is a reasonable reference model for most cases of similar joint action. The MSM does not give straight line isoboles and is suitable for situations where the constituents produce their effects in separate ways, neither influencing the effect of the other. It is a model for dissimilar joint action. She also emphasized that it does not mean that one of the models is more accurate than the other. They represent different hypotheses of which either or neither may be more appropriate in any given situation. (The original paper of Morse[1] may be consulted for detailed discussion of the two models and their working.)

III. OBSERVATIONS ON FISH

In light of the above discussion on the joint action of chemicals, the various reports on the combined toxicity of pesticides to fish may be examined. The toxicity of a second compound subsequent to exposure to a first compound, the toxicity of two or more compounds to which a fish is simultaneously exposed, and the methods of calculation of the joint toxicity are reviewed hereunder.

A. Pre-Exposure
There have been a number of reports describing the effect of pre-exposure on the toxicity of the same or a second pesticide to which the organism is exposed. It is not

clear, however, whether pre-exposure (also called preconditioning) confers any protection on the organism. Exposure of spot to 0.01 and 0.1 $\mu g/l$ toxaphene for 5 months made the fish more susceptible to the action of the same compound upon subsequent exposure, as evidenced by the increased acute toxicity of toxaphene (lowered LC 50 values) to pre-exposed fish in comparison with those that were not previously exposed.[6] Previous exposure of goldfish to alkyl benzene sulfonate (ABS) made them more susceptible to the toxic effects of dieldrin and DDT.[7] Higher mortality was recorded in the group that was pre-exposed to ABS. Cope,[8] quoting King,[9] reported that guppies, which were earlier exposed to low concentrations of DDT, were not adversely affected by exposure to DDT at 0.032 mg/l, a concentration that was highly toxic to fish that were not preconditioned.

Statham and Lech[10] exposed fingerling rainbow trout to a sublethal concentration of 2 mg/l carbaryl for 2 hr, and subsequently subjected them to this concentration of carbaryl along with either of two concentrations of a number of compounds (butyl ester of 2,4-D [2,4-dichlorophenoxyacetic acid], rotenone, PCP [pentachlorophenol], and dieldrin). They concluded that pre-exposure to carbaryl enhanced the toxicity of the other compounds. This conclusion is, however, based on the exposure of the fish to the different compounds (subsequent to the pre-exposure) only for 4 hr, an exposure too brief to draw any valid conclusions.

Pretreatment of rainbow trout with dieldrin resulted in reduced mortality upon subsequent exposure to DDT. The time until death, following the onset of visible symptoms of poisoning, was also prolonged by pretreatment with dieldrin. Further, the time necessary to reach the lethal levels of DDT in fish brain was also extended by the pretreatment.[11] On the contrary, pretreatment of the freshwater fish *Etheostoma nigrum* with dieldrin made the fish more susceptible to thermal stress.[12] The fish were exposed to a sublethal concentration of 2.3 $\mu g/l$ dieldrin for 30 days and the water temperature was increased at a rate of 1°C/day to a maximum of 7 to 9°C above the ambient temperature; the elevated temperature was maintained for 4 days and fish were maintained thereafter for a few weeks at that temperature to study the mortality and biochemical changes. Fish pretreated with dieldrin showed a mortality of 76.5% during the study period, as against only 13% in those fish that were not exposed to dieldrin, but only to the thermal stress. One interesting observation is that nearly half of the deaths in the pretreated fish occurred between 11 and 15 days, when there were many significant metabolic changes, too. Whole blood glucose levels peaked on the 10th day. Oxygen consumption of pretreated fish, which initially was less than that of controls until the 15th day, but increased thereafter, also was significantly different from that of the controls. The growth and feeding of pre-exposed fish was different from that of controls. While control fish had 92% of the initial feeding rate, the exposed fish had only 75%. The lipid content in pretreated fish was less than that of the controls, in the 1- to 15-day period. In the pretreated fish, pathological changes were evident (enlargement of liver, necrosis, hemorrhage, etc.). These results suggest that adaptation to low level chronic exposure to toxicants is achieved only at the cost of adjustibility to other low level environmental stress. Pre-exposure of rainbow trout fry to polychlorinated biphenyls (PCB) (Aroclor® 1254) for 30 days, to a concentration of 0.01 or 0.1 $\mu g/l$ made them more susceptible to the toxicity of rotenone and 2,4-D. On the other hand, because of the pre-exposure to PCB, the toxicity of malachite green was decreased. The toxicity of the pesticide, GD-174, was unaffected by pre-exposure to the higher concentration of PCB, but its toxicity was increased by the lower concentrations.[13]

The few studies on the pre-exposure of fish to various chemicals and the subsequent susceptibility of such fish to pesticides, do not permit any generalization. While in a few cases pre-exposure seemed to increase the mortality upon subsequent exposure to pesticides, in other instances the reverse was reported. Perhaps upon pre-exposure, activation of the detoxifying mechanisms like mixed function oxidase (MFO) (when induction of MFOs is possible; see Chapter 3, Volume 1), confers a certain amount of protection and results in decreased toxicity.

B. Observations on the Joint Toxicity of Pesticides to Fish

To test the possible interaction between methyl parathion and the cotton defoliant DEF® (S,S,S-tributyl phosphorotrithioate), *Gambusia* were exposed to 0.5 mg/l DEF®, or 5 mg/l methyl parathion, or both. There was no mortality of fish exposed to DEF® alone; 8% of the fish exposed to methyl parathion died, whereas there was 89% mortality of the fish exposed to both the toxicants. It was concluded that there was potentiation* of the toxicity of DEF® by methyl parathion. An important point that should not be overlooked, however, is that acetone at a concentration of 1 ml/l was used as a carrier solvent. To what extent this high concentration of acetone would have altered the membrane permeability and would have resulted in increased uptake of the toxicants, remains unknown.[14]

Ware[15] defined synergism as entomologists do, i.e., the toxicity of one compound is increased by another compound that is relatively or completely nontoxic. Quoting from the literature, he explained that EPN and malathion were synergistic in fish, whereas in the case of rats, this combination led to a marked potentiation. Apart from this apparent difference in the mode of action (it should be clear that the target site for the action of EPN and malathion is similar in both mammals and fish), neither EPN nor malathion can be considered nontoxic to fish and hence the resultant action, as explained, contradicts the definition of synergism, as given by Ware.

Simultaneous exposure of fathead minnows to parathion and linear alkyl benzene sulfonate (LAS) doubled the toxicity of parathion to the fish, the 96-h LC 50 of parathion, in the absence and presence of LAS being 1410 and 720 μg/l, respectively. The mean percentage of survival in 800 μg/l parathion solution was significantly reduced by LAS. Results of similar experiments with DDT were inconsistent, and between endrin and LAS there was no increase of toxicity. There was a significant difference in the amount of parathion bioconcentrated by fathead minnows surviving 48-hr exposure, in the presence or absence of 1 mg/l LAS. There was no difference in the amount of parathion taken up by the fish in the presence of lower concentrations of LAS. This suggests that at higher concentrations, the surfactant changed the membrane permeability. The conclusion of the authors that there appears to be no influence on the uptake of parathion by LAS because of inconsistent results in the other trials is untenable.[16] Enhanced joint toxicity of malathion and its alkaline hydrolysis product, dimethyl fumarate, was noted by Bender.[17]

Diquat was reported to be a potential inhibitor of aldrin epoxidation by the microsomal MFOs in the goldfish and mosquitofish. After mosquitofish were exposed to aldrin for 24 hr, dieldrin accounted for 23% of the residues extracted from the tissues. Addition of diquat to the test water at 8 × 10⁻⁴ M concentration led to the finding of negligible quantities of dieldrin in the water as well as hexane extracts of the tissues. The aldrin epoxidase system of various tissue homogenates of mosquitofish and goldfish was inhibited by diquat. The extent of inhibition was dose-dependent. The 24-hr toxicity of DDT, aldrin, and parathion was not affected by 275 mg/l diquat, whereas the acute toxicity of carbaryl was increased at this concentration.[18] The authors themselves pointed out that this finding is of little ecological significance, as such concen-

---

* In this section, the mode of action is described in the same terms as those used by the original authors, i.e., using the terms like "synergism", "potentiation", etc., even though it should be clear in the light of the remarks of Morse,[1] that these terms are untenable.

trations of diquat are never encountered in nature. Sesamex, present at about 1% concentration in the formulations of many organophosphate (OP) compounds and known to enhance their toxicity, was reported to decrease the extent of brain acetyl-cholinesterase (AchE) inhibition in golden shiners and green sunfish.

Exposure of rainbow trout orally to two levels of dieldrin (0.04 and 0.2 mg), DDT (0.2 and 1 mg), and methoxychlor (0.6 and 3 mg), on alternate days for 14 days, resulted in increased storage of DDT and methoxychlor in the presence of dieldrin. Storage of dieldrin was decreased in the presence of either DDT or methoxychlor. The storage of methoxychlor was decreased in the presence of DDT.[19]

The toxicity of many OP compounds to the fathead minnows was enhanced in the presence of the detergent LAS. A consistent relationship between molecular structure and enhanced toxicity was observed; the toxicity of those compounds with a substituted phenyl ring, with the exception of EPN and dicapthon, was enhanced in the presence of LAS.[20] Verma et al.[21] studied the joint toxicity of phenol, PCP, and dinitrophenol. Verma et al.[22] studied the joint toxicity of endosulfan, dichlorvos, and carbofuran to a freshwater fish. In both the studies, they extrapolated the dose-response curve and calculated the LC 0 and LC 100. As has been emphasized by Stephan,[23] to assume linearity near the extremes of the dose-response curve and extrapolation of the curve and calculated the LC 0 and LC 100. As has been emphasized by Stephan,[23] it is wrong to assume linearity near the extremes of the dose-response curve and extrapolation of the regression line that is useful for calculating the LC 50, to obtain LC 5 or LC 99, or some other value at the extremes, would be wrong.

Woodward[24] studied the toxicity of six paired mixtures of herbicides to the cutthroat trout. Picloram enhanced the toxicity of two formulations of 2,4-D. Dicamba® and isooctylester did not significantly alter the toxicity of other 2,4-D formulations or picloram.

Although endrin and methyl parathion were used together against cotton pests, methyl parathion had no effect on the toxicity of the former. Mount[25] stressed that when aquatic organisms are exposed to concentrations near the no-effect level, the question of increased toxicity of the mixtures does not arise.

C. Calculation of the Combined Toxicity of Pesticide Mixtures

Sun and Johnson[26] proposed that the theoretical toxicity of a mixture is equal to the sum of toxicity indices calculated from the percentage of each component multiplied by its respective toxicity index. The joint toxicity or co-toxicity coefficient of a mixture is calculated as follows:

$$\frac{\text{Actual toxicity index of a mixture (M)}}{\text{Theoretical toxicity index of a mixture}} \times 100$$

$$\text{Actual toxicity index of a mixture} = \frac{\text{LC 50 of A}}{\text{LC 50 of mixture}} \times 100$$

They explained that a coefficient nearer to 100 indicates similar action; a value < 100 indicates independent action, and a value > 100 indicates enhanced toxicity.

Gowing[27] and Colby[28] reviewed the various methods used to calculate the joint action of herbicide mixtures. Brown,[29] assuming that all poisons in a mixture contribute in a similar manner to the overall toxicity of a mixture, described a method to calculate the acute toxicity of mixtures. He proposed to estimate the proportion of the lethal concentration of a poison present in a mixture and then add the individual proportions. The proportional toxicity of each poison is obtained by dividing its concentration in water by the calculated LC 50 (48-h LC 50), i.e.,

Proportion of lethal concentration of A

$$= \frac{\text{concentration of toxicant A}}{\text{48-h LC 50 of A}}$$

Similarly, the proportions of the other toxicants in a mixture are calculated and are then summarized. If the value of this sum is < 1, it is considered that less than half the test organisms would die in a given time; if the value is > 1, more than half of the test organisms would die in that time. The main objection of this proposal is the assumption that all toxicants are equally toxic. Brown accepted that *a priori*, it is illogical to expect all poisons to contribute in the same extent to the overall toxicity of a mixture. Also, he discussed the limitations of this approach. Although the value of 1 is assigned as an arbitrary measure to define the toxicity of mixtures, the real value of the sum of proportional toxicity of mixtures may lie between 0.75 and 1.3.

The earlier approaches for evaluating the joint toxicity of pesticide mixtures, according to Macek,[30] have some shortcomings, like assumptions that may not be correct, statistical errors for toxicity indices not being developed satisfactorily, and rather involved algebraic manipulations. Instead of attempting to develop a mathematical model, Macek developed an empirical, algebraic evaluation to assess the acute toxicity of 29 two-chemical combinations to bluegill fish under static conditions. Toxicity tests were conducted simultaneously with concentration A_1, A_2, and A_3 of chemical A; B_1, B_2, and B_3 of chemical B; and A_1B_1, A_2B_2, and A_3B_3 of the combinations. Selecting concentrations of individual chemicals that produce $< 40\%$ mortality in 72 hr, the more-than-additive or less-than-additive toxicity index of combinations was calculated. The action was considered as additive when the ratio of the sum of mortalities produced at the various concentrations of AB to that produced by all levels of A plus that produced by all levels of B (AB/A+B) was approximately 1.0; the joint toxicity was considered as more-than-additive if the ratio was > 1.5, and as less-than-additive when the ratio was < 0.5. According to Macek, these estimates of trend are preliminary, and give an idea of mixtures that may be more or less toxic in combinations, than when used individually. Using this empirical approach, Macek concluded that out of 29 combinations tested, 11 had more-than-additive toxicity, 17 had additive toxicity, and 1 had less-than-additive toxicity. His results indicate that most of the OPs that were being proposed as alternatives to the long persistent organochlorines (OCs), had more-than-additive toxicity, when used in combinations. Malathion, one of the most frequently used compounds for control of aquatic pests, especially mosquitos, had more-than-additive toxicity with more than half of the compounds with which it was combined.

Marking[31] described a method in which individual toxic contributions of chemicals are summarized and the additive toxicity is defined by a linear index that expresses the toxicity quantitatively. The contributions of the components are summarized as follows:

$$(Am/Ai) + (Bm/Bi) = S$$

where A and B are chemicals,

i and m are toxicity (LC 50) of the

individual chemicals and mixtures

and S = sum of biological activity.

If the sum = 1, the toxicity is simply additive; if < 1, it indicates more-than-additive toxicity; > 1 indicates less-than-additive toxicity. Marking then proceeded to alter this system, by which the index represents additive, more-than-additive, and less-than-additive affects by zero, positive, and negative values, by assigning zero as a reference point for simple additive toxicity. Linearity is obtained by using the reciprocal values of S which are < 1, i.e., $(1/S)$-1. Then more-than-additive toxicity is represented by index values > 0.

Using the above method, Marking and Mauck[32] calculated the joint toxicity of of 20 pairs of insecticides used in the control of forest pests. Out of the 20, only two pairs showed only slightly more-than-additive toxicity, confirming the earlier observation of Mount.[25]

Calamari and Alabaster[33] attempted to explain the joint toxicity of mixtures, as Sprague did (additive, less-than, and more-than-additive toxicity), but their assumption that with two substances A and B, the "Effect of dose A = Effect of dose B" is open to question, for it is well known that two substances need not necessarily have the same effect, even on the same target organism.

Schaeffer et al.[34] assumed additivity of responses to mixtures of toxicants; the authors proposed to use multiple regression to study the joint action of mixtures of compounds. In this method, data obtained from the standard curves of substances and their mixtures are used. A t-test was devised, nonsignificant values of which support additivity, negative significant values indicate "antagonism", and positive significant values support "synergism".

Konemann,[35] while reviewing the various concepts of joint action in vogue, also noted that many definitions used in this field often lead to confusion. He was of the view that toxicity of mixtures of many chemicals can be theoretically predicted if only all of them act independently or if they have a simple similar action. Konemann[36] used a Mixture Toxicity Index (MTI) for comparing the results of toxicity of mixtures of chemicals. Using this method, Hermens and Leeuwangh[37] calculated the joint toxicity of five mixtures of 8 chemicals and another mixture of 24 chemicals.

The above review attempts to point out the inadequacy of the various proposals to calculate the joint action of pesticide mixtures on fish. Although reviews like that of Morse[1] have helped clear the picture of the joint action of herbicides, and can perhaps be extended to the subject under discussion, too, such an attempt is yet to be made, on the basis of well-planned experimental work, regarding the toxicity of pesticide mixtures to fish.

REFERENCES

1. Morse, P. M., Some comments on the assessment of joint action in herbicide mixtures, *Weed Sci.*, 26, 58, 1978.
2. Sprague, J. B., Measurement of pollutant toxicity to fish. II. Utilizing and applying bioassay results, *Water Res.*, 4, 3, 1970.
3. Gaddum, J. H., *Pharmacology*, 3rd ed., Oxford University Press, London, 1948.
4. Schubert, J., Riley, E. J., and Tyler, S. A., Combined effects in toxicology — a rapid systematic testing procedure: cadmium, mercury, and lead, *J. Toxicol. Environ. Health*, 4, 763, 1978.
5. Akobundu, I. O., Sweet, R. D., and Duke, W. B., A method of evaluating herbicide combinations and determining herbicide synergism, *Weed Sci.*, 23, 20, 1975.
6. Lowe, J. J., Chronic exposure of spot, *Leiostomus xanthurus*, to sublethal concentrations of toxaphene in sea water, *Trans. Am. Fish. Soc.*, 93, 396, 1964.
7. Dugan, P. R., Influence of chronic exposure to anionic detergents on toxicity of pesticides to goldfish, *J. Water Pollut. Control Fed.*, 39, 63, 1967.
8. Cope, O. B., Contamination of the freshwater ecosystem by pesticides, *J. Appl. Ecol.*, 3, 33, 1966.
9. King, S. F., Some effect of DDT on the guppy and the brown trout, *Fish Wildl. Serv., Bur. Sport Fish. Wildl., Spec. Sci. Rep. Fish No.* 399, 1962.

10. Statham, C. N. and Lech, J. J., Potentiation of the acute toxicity of several pesticides and herbicides in trout by carbaryl, *Toxicol. Appl. Pharmacol.*, 34, 83, 1975.

11. Mayer, F. L., Jr., Street, J. C., and Neuhold, J. M., DDT intoxication in rainbow trout as affected by dieldrin, *Toxicol. Appl. Pharmacol.*, 22, 347, 1972.

12. Silbergeld, E. K., Dieldrin, effects of chronic sublethal exposure on adaptation to thermal stress in freshwater fish, *Environ. Sci. Technol.*, 7, 846, 1973.

13. Bills, T. D., Marking, L. L., and Mauck, W. L., Polychlorinated biphenyl (Aroclor 1254) residues in rainbow trout: effects on sensitivity to nine fishery chemicals, *N. Am. J. Fish. Manage.*, 1, 200, 1981.

14. Fabacher, D. L., Davis, J. D., and Fabacher, D. A., Apparent potentiation of the cotton defoliant DEF by methyl parathion in mosquitofish, *Bull. Environ. Contam. Toxicol.*, 16, 716, 1976.

15. Ware, G. W., Effects of pesticides on nontarget organisms, *Residue Rev.*, 76, 173, 1980.

16. Solon, J. M., Lincer, J. L., and Nair, J. H., III, The effect of sublethal concentration of LAS on the acute toxicity of various insecticides to the fathead minnow (*Pimephales promelas* Rafinesque), *Water Res.*, 3, 767, 1969.

17. Bender, M. E., The toxicity of the hydrolysis and breakdown products of malathion to the fathead minnow (*Pimephales promelas* Rafinesque), *Water Res.*, 3, 571, 1969.

18. Krieger, R. I. and Lee, P. W., Inhibition of *in vivo* and *in vitro* epoxidation of aldrin, and potentiation of toxicity of various insecticide chemicals by diquat in two species of fish, *Arch. Environ. Contam. Toxicol.*, 1, 112, 1973.

19. Mayer, F. L., Jr., Street, J. C., and Neuhold, J. M., Organochlorine insecticide interactions affecting residue storage in rainbow trout, *Bull. Environ. Contam. Toxicol.*, 5, 300, 1970.

20. Solon, J. M. and Nair, J. H., III, The effect of a sublethal concentration of LAS on the acute toxicity of various phosphate pesticides to the fathead minnow (*Pimephales promelas* Rafinesque), *Bull. Environ. Contam. Toxicol.*, 5, 408, 1970.

21. Verma, S. R., Rani, S., and Dalela, R. C., Synergism, antagonism and additivity of phenol, pentachlorophenol, and dinitrophenol to a fish (*Notopterus notopterus*), *Arch. Environ. Contam. Toxicol.*, 10, 365, 1981.

22. Verma, S. R., Rani, S., Bansal, S. K., and Dalela, R. C., Effects of the pesticides Thiotox, dichlorovos and carbofuran on the test fish *Mystus vittatus*, *Water Air Soil Pollut.*, 13, 229, 1980.

23. Stephan, C. E., Methods for calculating an LC$_{50}$, *Aquatic Toxicology and Hazard Evaluation*, ASTM STP 634, Mayer, F. L. and Hamelink, J. L., Eds., American Society for Testing and Materials, Philadelphia, 1977, 65.

24. Woodward, D. F., Acute toxicity of mixtures of range management herbicides to cutthroat trout, *J. Range Manage.*, 35, 539, 1982.

25. Mount, D. I., Adequacy of laboratory data for protecting aquatic communities, in *Analyzing and Hazard Evaluation Process*, Dickson, K. L., Maki, A. W., and Cairns, J., Jr., Eds., American Fisheries Society, Washington, D.C., 1979, 112.

26. Sun, Y. P. and Johnson, E. R., Analysis of joint action of insecticides against house flies, *J. Econ. Entomol.*, 53, 887, 1960.

27. Gowing, D. P., Comments on tests of herbicide mixtures, *Weeds*, 8, 379, 1960.

28. Colby, S. R., Calculating synergistic and antagonistic responses of herbicide combinations, *Weeds*, 15, 20, 1967.

29. Brown, V. M., The calculation of the acute toxicity of mixtures of poisons to rainbow trout, *Water Res.*, 2, 723, 1968.

30. Macek, K. J., Acute toxicity of pesticide mixtures to bluegills, *Bull. Environ. Contam. Toxicol.*, 14, 648, 1975.

31. Marking, L. L., Method for assessing additive toxicity of chemical mixtures, *Aquatic Toxicology and Hazard Evaluation*, ASTM STP 634, Mayer, F. L. and Hamelink, J. L, Eds., American Society for Testing and Materials, Philadelphia, 1977, 99.

32. Marking, L. L. and Mauck, W. L., Toxicity of paired mixtures of candidate forest insecticides to rainbow trout, *Bull. Environ. Contam. Toxicol.*, 13, 518, 1975.

33. Calamari, D. and Alabaster, J. S., An approach to theoretical models in evaluating the effects of mixtures of toxicants in the aquatic environment, *Chemosphere*, 9, 533, 1980.

34. Schaeffer, D. J., Glave, W. R., and Janardan, K. G., Multivariate statistical methods in toxicology. III. Specifying joint toxic interaction using multiple regression analysis, *J. Toxicol. Environ. Health*, 9, 705, 1982.

35. Könemann, H., Structure-activity relationships and additivity in fish toxicities of environmental pollutants, *Ecotoxicol. Environ. Saf.*, 4, 415, 1980.

36. Konemann, H., Fish toxicity tests with mixtures of more than two chemicals: a proposal for a quantitative approach and experimental results, *Toxicology*, 19, 229, 1976.

37. Hermens, J. and Leeuwangh, P., Joint toxicity of mixtures of 8 and 24 chemicals to the guppy (*Poecilia reticulate*), *Ecotoxicol. Environ. Saf.*, 6, 302, 1982.

Chapter 5

INSECTICIDE RESISTANCE IN FISH

I. INTRODUCTION

It is said that while modern, organic pesticides are man's answer to pest problems, resistance is the strategy of the insect to the intense and indiscriminate use of pesticides. In the early part of this century, occurrence of insect populations resistant to inorganicals was sporadic, but with the advent of the era of synthetic organicals, many insects developed resistance to several different compounds.

Among vertebrates, tolerance of organic pesticides is not as common as has been reported in the case of insects. The earliest record of the tolerance of fish to an insecticide was made by Vinson et al.,[1] who reported that mosquitofish, bred in areas with a long history of pesticide use, were resistant to DDT. Since then, many more resistant populations of different species of fish have come to light. It is remarkable that most of the resistant populations on record have been reported from a single area in the state of Mississippi, and most of the investigations on the resistance of fish have come from a single institution, i.e., the Mississippi State University, State College, Miss.

A. Resistance to Different Compounds

Drainage canals and ditches, located in the intensely sprayed cotton fields in the Belzoni area of Mississippi, furnished many of the different populations of resistant fish that have been regularly studied over a number of years. The heavy spraying of the adjacent cotton fields with a variety of insecticides (DDT, toxaphene, endrin, and organophosphate [OP] compounds like methyl parathion, etc.) at different times, provided the necessary selection pressure for the elimination of the susceptible members of the population and the survival and establishment of the resistant ones.

Among insects, basically two types of insecticide resistance are known, viz., resistance to DDT and resistance to cyclodienes. Within each type, cross-resistance extends to structurally similar compounds. Thus, a population that is resistant to DDT may show a certain amount of tolerance to methoxychlor, DDD, and other closely related compounds; likewise, resistance to endrin may extend to resistance to dieldrin or heptachlor. This is known as cross-resistance. Insecticide resistance in fish is somewhat akin to that reported in soil insects, where resistance to DDT has been somewhat low, when compared with resistance to cyclodienes.[2] Among the fish, although resistance to DDT was the first to be reported, it soon became clear that mosquitofish from the Belzoni area displayed a greater degree of tolerance to the cyclodienes and the chlorinated camphane, toxaphene, rather than to the DDT type compounds.

Comparison of the 48-h LC 50 values of 28 pesticides to resistant (R) population of *Gambusia* from the Belzoni area with those to a susceptible (S) population from Mississippi State showed that the toxicity of the cyclodiene compounds to the R-fish was 54 (dieldrin) to 568 (heptachlor) times less.[3] The difference in the toxicity of DDT group compounds to the R- and S-populations was much narrower, ranging from 1.7 for methoxychlor to 5.1 for DDT. Of the several compounds for which high resistance was evident, only two compounds, viz., toxaphene and endrin, were regularly sprayed for 4 years preceding this investigation. The increased resistance to Strobane®, heptachlor, and aldrin recorded seems to be the result of cross-resistance. Furthermore, little resistance was developed to certain other compounds like methyl parathion, despite their extensive use in the Belzoni area.

The possible tolerance of three herbicides by R-variety mosquitofish was tested. No difference in the toxicity of DBNP to the S- and R-mosquitofish was noticed, whereas 2,4-D (2,4-dichlorophenoxyacetic acid) butyl ester and trifluralin were 1.87 and 2.05 times less toxic to the R-fish.[4]

Other species of fish like golden shiners, green sunfish, and bluegills from the same area as in the above investigation were reported to the endrin-resistant.[5] Subsequently, endrin-resistant yellow bullheads were also recorded from the Belzoni area.[6] Different species seem to tolerate the same compound to different degrees. While mosquitofish could tolerate a 36-hr exposure to 1500 μg/ℓ endrin, and golden shiners 1000 μg/ℓ endrin, the 36-h LC 50 for yellow bullheads (R-population) was 75 μg/ℓ.[6] It was suggested that bottom dwelling fish have little access to pesticide-laden insects and hence are exposed to less amounts of the pesticide in water. Although they are more tolerant than the S-populations of the same species from uncontaminated areas, R-populations of certain (deep dwelling) species have less tolerance than other species that are exposed to a greater quantity of the pesticide, in the same contaminated area.

Finley et al.[7] investigated the possible selection pressures that could lead to the establishment of R-populations. Feeding experiments (R-mosquitofish were fed to S- and R-green sunfish) showed that the selection pressure via the food chain was minimal compared with direct exposure, as a result of runoff enrichment of residues, following rains.

The selection pressures that lead to the establishment of R-populations and the genetic plasticity of such "opportunistic species" have been reviewed.[8] Since resistance is the result of a selection pressure exerted on a population until the genes that express increased tolerance are fairly wellspread in the population, the establishment of a R-variety obviously requires the time span of several generations. On the other hand, Cope,[9] quoting the work of King,[10] stated that resistance to DDT by fish was noticeable within a short period of exposure. It was reported that when guppies were preconditioned by exposure to extremely low concentrations of DDT, subsequent exposure to a DDT concentration of 32 μg/ℓ (previously known to be lethal to normal fish) resulted in no mortality. Upon being maintained in the laboratory for some time, the R-fish lost some of the tolerance to parathion, and at the end of 3 weeks the toxicity of parathion to both R- and S- varieties was the same.[11]

R-populations of mosquitofish were also reported from two sites near College Station, Tex.[12] Unlike the Mississippi population, the Texas populations were more tolerant of the DDT group than the cyclodiene group. Tolerance of the cyclodiene group, which was 100 to 500 times that of the susceptible population in the Mississippi area, was a mere 4 to 10 times in the Texas area. Another area near College Station, Tex., draining urban lands, also had a population of mosquitofish with a low level resistance to DDT and cyclodienes. Further, the highly DDT-resistant population of mosquitofish had a higher Σ DDT concentration (50 mg/kg) than the other two populations from Texas — one less resistant and the other a susceptible strain. Although resistance to toxaphene and aldrin was evident, GLC analysis revealed no residues of either. It was postulated that the resistance to these two compounds was inherited from an ancestral stock, which would have been repeatedly exposed to these residues.[12]

When S- and R-fish were exposed to 10.2 μg/ℓ ^{14}C-endrin for 8 hr, the former had 13.5 times higher residue concentration in the gills; maximum difference in the tissue concentration was found in the brain, suggesting the existence of a membrane barrier in the R-fish, as a protective measure. Further, R-fish could metabolize and excrete endrin more effectively than the S-fish.

B. Relative Toxicity of Pesticides to Resistant (R) and Susceptible (S) Populations

In the Belzoni area, DDT and methyl parathion were relatively more toxic to the R-fish than to the S-fish, whereas with endrin and toxaphene the reverse was true.[13] It was found that despite a 1000-fold difference in the 36-h LC 50 value of endrin to R- and S-varieties, both populations removed endrin at the same rate from static test solutions.[14] While mortality of S-test fish (from very low to very high percent) was caused by a narrow range of the toxicant, with R-fish the mortality increased gradually over a wide range. Both varieties were more sensitive to endrin in late summer than in the fall and winter.[14] R-fish, collected in the field, contained 6.8 to 12 mg/kg endrin in late August but only 0.73 to 0.88 mg/kg in late February, indicating that during spring and summer, they accumulated higher quantity of residues owing to continuous exposure to the contaminant in the agricultural runoff. No endrin residues could be detected in the S-variety, either in summer or in winter. The toxicity of parathion was less in spring than in fall for both S- and R-populations of mosquitofish.[11]

R-varieties are a potential population hazard. One group of R-mosquitofish was reported to have survived with a residue concentration of 214.3 mg/kg[16] and such fish pose a grave risk to the predators. This is perhaps the reason why top carnivores like largemouth bass were absent in localities supporting R-mosquitofish populations.[7] A single R-fish, upon transfer from 1 mg/l endrin solution to clean tapwater, released enough endrin to kill five S-fish.[14] Not only do live R-fish excrete enough quantity of the residue to kill normal fish, they do so even when dead.[14]

When both the S- and R-*Gambusia* were exposed to the same quantity of endrin for the same period, the brain and liver of dead S-fish contained 1.9 times more endrin than live R-fish.[15] This makes it clear that the uptake of endrin by R-fish was somewhat slower than by S-fish and is contrary to the earlier report that S- and R-fish removed endrin from aqueous solutions at the same rate[14] under static conditions. The ratio of the residues in brain to liver was the same in both dead S- and live R-fish.[15] Similarly, when both R- and S-fish were killed by exposure to 1000 μg/l endrin, the brains of S-fish had more endrin than those of R-fish. Interestingly, at the time of death, the livers of R-fish had more endrin than the livers of S-fish in this experiment, the respective values (mg/kg) in S- and R-fish being, brain 149.3 and 57.5, and liver 160 and 353. The S-fish were killed in 45 min and R-fish in 2 hr. At the time of death, the residue concentration in the brain was about 92% of that of the liver in S-fish, whereas it was about 16% in the R-fish. In the R-fish, the rate of uptake was not only slower, but much of the residue was accumulated in the liver, indicating an efficient blood-brain barrier. The slower uptake by fish and slower entry into the target site (brain) explain the greater tolerance of insecticides in R-fish.[15]

C. Biochemical Basis of Resistance

Differences in (1) the size of the lipid pool, (2) rate of uptake, (3) efficiency of binding to the membranes, and (4) rate of detoxification were investigated to explain the increased tolerance of R-fish to insecticides.

The lipid content of R-liver homogenates was higher than that of the S-liver homogenates.[16] The lipid content of R-fish was 1.8 times that of S-fish. When the R-fish were removed and kept in uncontaminated ponds for 6 months, the lipid content of such fish was only 1.2 times that of S-fish. Similarly, the lipid concentration of the livers of R-fish was 1.7 and 1.1 times that of the livers of S-fish before and after the maintenance of R-fish in uncontaminated ponds for 6 months.[17] It was suggested that the increased lipid content of the resistant variety helped sequester much of the toxicant load and avoid the toxic effect. The livers of R-fish were significantly larger. Fabacher and Chambers[17] also stressed that increased lipid content only partially explains the increased tolerance to endrin and toxaphene, since such fish did not have an increased tolerance to other lipid soluble organochlorine (OC) compounds like DDT. They also suggested that binding to nonessential proteins in R-fish may also contribute to the

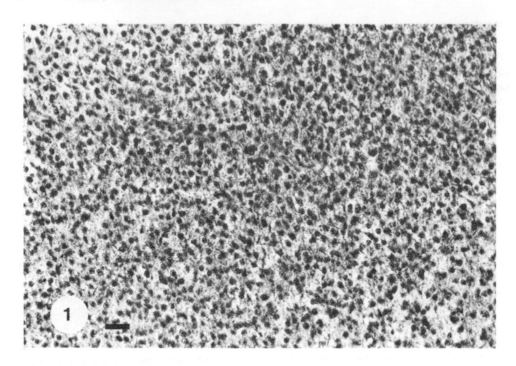

FIGURE 1.   Photomicrograph of liver cells from insecticide-susceptible *Gambusia affinis.* Hematoxylin and eosin. Line marker = 20 μm. (From Yarbrough, J. D. and Coons, L. B., *Chem. Biol. Interact.,* 10, 247, 1975. With permission.)

tolerance. Myelin, isolated from S- and R-mosquitofish treated with ^{14}C-aldrin or ^{14}C-dieldrin, showed greater retention of either of the compounds by S-fish.[18] Structural alteration of the myelin in R-fish seems to have caused less retention and hence, reduced uptake of the toxicants. Electron microscopy revealed a doubling of the myelin sheaths in R-fish tissue preparations. The blood-brain barrier was once again confirmed by Wells and Yarbrough,[18] with lesser amounts of dieldrin being present in the brains of R-fish in comparison with the livers, unlike in S-fish.

Some cytological differentiation between the two strains was evident. The R-fish had enlarged liver cells. On an average, the hepatocytes of an R-fish were 1.75 times larger than those of an S-fish (Figures 1 and 2). A marked increase in the lipid inclusions of the hepatocytes of R-fish was noticed (Figures 3 and 4). No difference in the other cellular components like the smooth and rough endoplasmic reticulum, glycogen granules, etc. was evident between the two races of fish.[19]

In vitro studies on the binding of ^{14}C-endrin to the cellular fractions of S- and R-mosquitofish revealed that cell membrane fractions of R-fish bound more endrin and as a consequence less of the toxicant entered the cell. In vivo exposure of S- and R-fish to 2 μg/ℓ endrin confirmed the existence of a blood-brain barrier in the R-fish; significantly higher levels of endrin were present in the brain of S-fish than in the tolerant group. Further, the S-fish had higher residue concentration in the brain relative to the liver, unlike in the R-fish, the liver to brain ratio being 3:1 in S-fish and 7.7 to 1 in the R-fish.[20] Structural differences in the myelin were also demonstrable which could also perhaps account for the differences in the sensitivity of the two strains. The R-cell membrane could bind twice as much endrin as the S-cell membrane.

Apart from the membrane barrier, the decreased sensitivity of R-fish to insecticides is also partly based on their ability to metabolize the toxicant rapidly. Although the

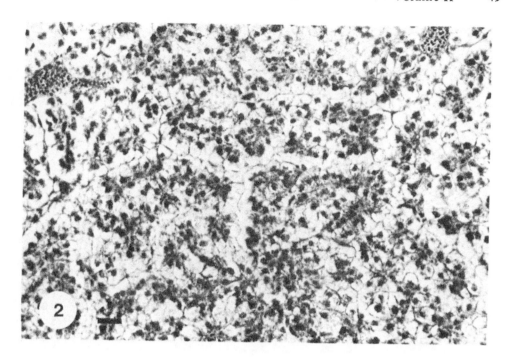

FIGURE 2. Photomicrograph of liver cells from insecticide-resistant *Gambusia affinis*. Hematoxylın and eosin. Line marker = 20 μm. (From Yarbrough, J. D. and Coons, L. B , *Chem. Biol. Interact.*, 10, 247, 1975. With permission.)

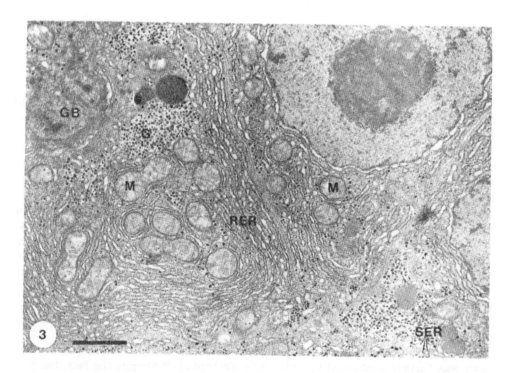

FIGURE 3. Electron micrograph of ınsecticide-susceptible *Gambusıa affinis* hepatocytes. SER, smooth endoplasmic retıculum; RER, rough endoplasmıc retıculum; G, glycogen; GB, golgı body; M, mitochondria. Lıne marker = 1 μm. (From Yarbrough, J. D. and Coons, L. B., *Chem Biol. Interact.*, 10, 247, 1975. With permıssion.)

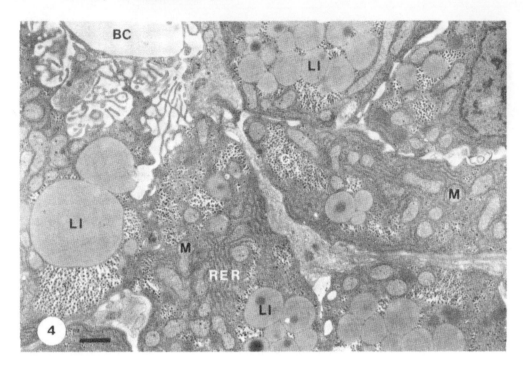

FIGURE 4.   Electron micrograph of insecticide-resistant *Gambusia affinis* hepatocytes. BC, bile caniculi;
LI, lipid inclusions; rest of abbreviations as in Figure 3. Line marker = 1 μm. (From Yarbrough, J. D. and
Coons, L. B., *Chem. Biol. Interact.*, 10, 247, 1975 With permission.)

uptake of ^{14}C-aldrin by both R- and S-strains of mosquitofish was comparable, the
former converted aldrin to dieldrin, and water soluble components, to a greater extent
than the latter.[21] Significantly more dieldrin was present in the brain of S-fish than in
the brain of R-fish, and in the livers of R-fish than in the livers of S-fish.

R-mosquitofish (resistant to cyclodienes and toxaphene) also showed a small degree
of increased tolerance to parathion and methyl parathion. The R-population had a
higher activity of methyl parathion degrading enzymes. Increased tolerance of R-fish
to parathion seems to be the result of increased mixed function oxidase activity, leading
to increased detoxification.[22] Higher levels of mixed function oxidase (MFO), which
dearylated parathion or methyl parathion, were the cause of greater tolerance of these
two compounds in R-fish.

S-mosquitofish, both those that showed symptoms of poisoning and those that did
not, absorbed significantly higher quantities of aldrin than did the R-fish exposed to
the same concentration. No such difference in the uptake of dieldrin was noticeable,
perhaps because of the greater polarity of dieldrin. Slower uptake and a more efficient
blood-brain barrier in the R-fish coupled with a more effective metabolism and excretion explain the greater tolerance of R-fish to aldrin.[23] In S-fish, a direct correlation
between the whole body dieldrin concentration and the appearance of toxic symptoms
existed; however, no apparent relationship between tissue insecticide concentration and
the onset of toxic symptoms could be found.

The existence of an effective barrier for the penetration of the toxicant at the target
organ was further confirmed by Scales and Yarbrough.[24] Despite the fact that S-fish
were exposed to 10 μg/*l* endrin and R-fish to 1500 μg/*l* endrin, the amount of endrin
present in both groups at the time of appearance of the symptoms was reported to be

the same. In the R-group, the fish that were showing symptoms of poisoning, had higher residue concentration than those that did not manifest any symptoms of poisoning. On the other hand, S-fish that did not show any symptoms had higher residue concentration than those showing symptoms of poisoning. This indicates that endrin resistance in mosquitofish is due to the presence of a less sensitive target site in the more tolerant fish.

Although rotenone, a fish poison of plant origin, was never used in the Belzoni area, the resistant populations of mosquitofish were found to be resistant to rotenone, too.[25] Rotenone resistance is presumed to be due to cross-resistance, and functionally it was the result of increased levels of MFO enzymes, that could oxidize and eliminate rotenone.

OC-resistant mosquitofish showed a 3.4-fold tolerance to pyrethrum, when compared with the LC 50 value to the OC-susceptible strain.[26] The tolerance to pyrethrum was only partly but not wholly the result of incresed MFO activity. R-fish that were maintained in insecticide-free environment for several generations, also showed resistance to pyrethrum, though to a decreased extent, emphasizing the genetic control of pyrethrum resistance.

Seasonal variation in the mixed function oxidase components in the S- and R-strains of mosquitofish was investigated by Chambers and Yarbrough.[27] Many of the MFO factors studied, like cytochrome p-450 and b_5, NADPH-cytochrome C, and NADPH-cytochrome b_5 reductase, as well as microsomal protein and ratio of liver to body weight, showed highest values during late fall and winter. The R-strain had significantly larger livers and therefore greater amounts of microsomal components than the S-strain.

The R-population possessed higher peripheral acetylcholinesterase (AChE) and carboxylesterase activity than the S-population.[28] The higher levels of carboxylesterases, because of their greater affinity for OPs, bind the inhibitor molecules to a greater extent and contribute to a higher OP tolerance in the R-fish. Lower levels of liver esterases in the R-fish and hence, reduced levels of hydrolysis of 2,4-D in the liver, led to a slight increase in the tolerance of the insecticide by R-fish.[29]

In vitro exposure of intact brain mitochondrial preparations from R-fish to 10^{-7} to 10^{-4} M endrin had no effect on the succinic dehydrogenase (SDH) activity.[16] A similar exposure of the mitochondrial preparations from S-fish, however, showed inhibition of SDH activity. When the mitochondrial membrane was disrupted by repeated freezing and thawing, SDH activity was inhibited in the mitochondrial preparations of both S- and R-fish. Similar results were obtained with DDT, dieldrin, and toxaphene,[30] confirming the existence of a membrane barrier in R-mosquitofish.

Using intact and disrupted mitochondria, the sensitivity and resistance of the mosquitofish to mirex was investigated. Increased SDH activity in disrupted mitochondrial preparations of S- and R-populations of mosquitofish (except in the case of brain mitochondrial preparation of S-fish) was observed. There was stimulation at 10^{-4} M concentration of mirex, whereas at a concentration of 10^{-6} M, SDH activity was inhibited. The stimulation at 10^{-4} M may be the result of increased membrane permeability.[31]

The cell membrane from the brain and liver preparations of S-fish retained less endrin than similar preparations from R-fish, but the reverse was true with the mitochondrial membrane.[32] Cell membranes of R-fish retained more DDT than did S-fish cell membranes.[33] In general, it appears that the cell membranes in R-fish either prevent or slow down the insecticide uptake, which results in the low levels of insecticides in the organs of R-fish.

Although R-fish are known to tolerate high concentrations of certain insecticides, they are more susceptible to oxygen stress than S-fish. When compared with an LC 50 of 131.3 mg/ℓ of formaldehyde under nonaerated conditions (leading to poor oxygen concentrations) to the S-fish, the LC 50 value of formaldehyde under similar conditions to R-fish was 56.6 mg/ℓ.[34] McCorkle et al.[34] emphasized that "Although the resistant population has been selected by environmental pressures for enhanced tolerance to pesticides, the selection process appears to have resulted in a simultaneous decrease in the population's tolerance of other environmental stresses."

In conclusion, the extensive studies of Yarbrough and associates indicate that insecticide resistance in fish is operative because of a protective membrane barrier at the site of action and also a decrease in the sensitivity of the target site.[32] This barrier is present in both the populations of mosquitofish, but is more effective in the R-fish.

REFERENCES

1. Vinson, S. B., Boyd, C. E., and Ferguson, D. E., Resistance to DDT in the mosquitofish, *Gambusia affinis*, *Science*, 139, 217, 1963.
2. Harris, C. R., Insecticide resistance in soil insects attacking crops, in *Pesticide Management and Insecticide Resistance*, Watson, D. L. and Brown, A. W. A., Eds., Academic Press, New York, 1977, 321.
3. Culley, D. D., Jr. and Ferguson, D. E., Patterns of insecticide resistance in the mosquitofish, *Gambusia affinis*, *J. Fish. Res. Board Can.*, 26, 2395, 1969.
4. Fabacher, D. L. and Chambers, H., Resistance to herbicides in insecticide-resistant mosquitofish, *Gambusia affinis*, *Environ. Lett.*, 7, 15, 1974.
5. Ferguson, D. E., Culley, D. D., Cotton, W. D., and Dodds, R. P., Resistance to chlorinated hydrocarbon insecticides in three species of freshwater fish, *BioScience*, 14, 43, 1964.
6. Ferguson, D. E. and Bingham, C. R., Endrin resistance in the yellow bullhead, *Ictalurus natalis*, *Trans. Am. Fish. Soc.*, 95, 325, 1966.
7. Finley, M. T., Ferguson, D. E., and Ludke, J. L., Possible selective mechanisms in the development of insecticide-resistant fish, *Pestic. Monit. J.*, 3, 212, 1970.
8. Luoma, S. N., Detection of trace contaminant effects in aquatic ecosystems, *J. Fish. Res. Board Can.*, 34, 436, 1977.
9. Cope, O. B., Contamination of the freshwater ecosystem by pesticides, *J. Appl. Ecol.*, 3, 33, 1966.
10. King, S. F., Some effects of DDT on the guppy and the brown trout, *Fish Wildl. Serv., Bur. Sport Fish. Wildl., Spec. Sci. Rep Fish. No. 399*, 1962.
11. Chambers, J. E. and Yarbrough, J. D., Parathion and methyl parathion toxicity to insecticide-resistant and susceptible mosquitofish *(Gambusia affinis)*, *Bull. Environ. Contam. Toxicol.*, 11, 315, 1974.
12. Dziuk, L. J. and Plapp, F. W., Insecticide resistance in mosquitofish from Texas, *Bull. Environ. Contam. Toxicol.*, 9, 15, 1973.
13. Ferguson, D. E. and Bingham, C. R., The effects of combinations of insecticides on susceptible and resistant mosquitofish, *Bull. Environ. Contam. Toxicol.*, 1, 97, 1966.
14. Ferguson, D. E., Ludke, J. L., and Murphy, G. G., Dynamics of endrin uptake and release by resistant and susceptible strains of mosquitofish, *Trans. Am. Fish. Soc.*, 95, 335, 1966.
15. Fabacher, D. L. and Chambers, H., Uptake and storage of ^{14}C-labeled endrin by the livers and brains of pesticide-susceptible and resistant mosquitofish, *Bull. Environ. Contam. Toxicol.*, 16, 203, 1976.
16. Yarbrough, J. D. and Wells, M. R., Vertebrate insecticide resistance: the *in vitro* endrin effect on succinic dehydrogenase activity on endrin-resistant and susceptible mosquitofish, *Bull. Environ. Contam. Toxicol.*, 6, 171, 1971.
17. Fabacher, D. L. and Chambers, H., A possible mechanism of insecticide resistance in mosquitofish, *Bull. Environ. Contam. Toxicol.*, 6, 372, 1971.
18. Wells, M. R. and Yarbrough, J. D., *In vivo* and *in vitro* retention of (^{14}C) aldrin and (^{14}C) dieldrin in cellular fraction from brain and liver tissues of insecticide-resistant and susceptible *Gambusia*, *Toxicol. Appl. Pharmacol.*, 24, 190, 1973.
19. Yarbrough, J. D. and Coons, L. B., A comparative cytological study between hepatocytes of insecticide-resistant and susceptible mosquitofish *(Gambusia affinis)*, *Chem. Biol. Interact.*, 10, 247, 1975.
20. Wells, M. R. and Yarbrough, J. D., Vertebrate insecticide resistance: *in vivo* and *in vitro* endrin binding to cellular fractions from brain and liver tissues of *Gambusia*, *J. Agric. Food Chem.*, 20, 14, 1972.

21. Wells, M. R., Ludke, J. L., and Yarbrough, J. D., Epoxidation and fate of (^{14}C) aldrin in insecticide-resistant and susceptible populations of mosquitofish *(Gambusia affinis)*, *J. Agric. Food Chem.*, 21, 428, 1973.

22. Chambers, J. E. and Yarbrough, J. D., Organophosphate degradation by insecticide-resistant and susceptible populations of mosquitofish *(Gambusia affinis)*, *Pestic. Biochem. Physiol.*, 3, 312, 1973.

23. Watkins, J. and Yarbrough, J. D., Aldrin and dieldrin uptake in insecticide-resistant and susceptible mosquitofish *(Gambusia affinis)*, *Bull. Environ. Contam. Toxicol.*, 14, 731, 1975.

24. Scales, E. H. and Yarbrough, J. D., Endrin uptake in insecticide-resistant and susceptible mosquitofish *(Gambusia affinis)*, *J. Agric. Food Chem.*, 23, 1076, 1975.

25. Fabacher, D. L. and Chambers, H., Rotenone tolerance in mosquitofish, *Environ. Pollut.*, 3, 139, 1972.

26. Fabacher, D. L. and Chambers, H., Apparent resistance to pyrethroids in organochlorine-resistant mosquitofish, *Proc. 26th Annu. Conf. Southeastern Assoc. Game Fish Comm.*, 1972, 461.

27. Chambers, J. E. and Yarbrough, J. D., A seasonal study of microsomal mixed-function oxidases components in insecticide-resistant and susceptible mosquitofish, *Gambusia affinis*, *Toxicol. Appl. Pharmacol.*, 48, 497, 1979.

28. Chambers, J. E., The relationship of esterases to organophosphorus insecticide tolerance in mosquitofish, *Pestic. Biochem. Physiol.*, 6, 517, 1976.

29. Chambers, H., Bziuk, L. J., and Watkins, J., Hydrolytic activation and detoxification of 2,4-dichlorophenoxyacetic acid esters in mosquitofish, *Pestic. Biochem. Physiol.*, 7, 297, 1977.

30. Moffett, G. B. and Yarbrough, J. D., The effects of DDT, toxaphene, and dieldrin on succinic dehydrogenase activity in insecticide-resistant and susceptible *Gambusia affinis*, *Agric. Food Chem.*, 20, 588, 1972.

31. McCorkle, F. M. and Yarbrough, J. D., The *in vitro* effects of mirex on succinic dehydrogenase activity in *Gambusia affinis* and *Lepomis cyanellus*, *Bull. Environ. Contam. Toxicol.*, 11, 364, 1974.

32. Yarbrough, J. D., Insecticide resistance in vertebrates, in *Survival in Toxic Environments*, Khan, M. A. Q. and Bederka, J. P., Eds., Academic Press, New York, 1974, 373.

33. Wells, M. R. and Yarbrough, J. D., Retention of ^{14}C-DDT in cellular fractions of vertebrate insecticide-resistant and susceptible fish, *Toxicol. Appl. Pharmacol.*, 22, 409, 1972.

34. McCorkle, F. M., Chambers, J. E., and Yarbrough, J. D., Tolerance of low oxygen stress in insecticide-resistant and susceptible populations of mosquitofish *(Gambusia affinis)*, *Life Sci.*, 25, 1513, 1979.

Chapter 6

SUBLETHAL EFFECTS OF PESTICIDES ON FISH

I. INTRODUCTION

Mass mortality of fish due to pesticide exposure is rare, and results only from accidents or direct spraying of the water bodies. More commonly, fish are subjected to long-term stress arising from exposure to sublethal concentrations. In the long run, these sublethal concentrations may prove more deleterious than the lethal concentrations, because subtle and small effects on the fish may alter their behavior, feeding habits, position in the school, reproductive success, etc. Behavioral or morphological changes may make the fish more conspicuous in the environment and more susceptible to predation or parasitization, thereby reducing the ability of the population to survive and reproduce. Likewise, subtle effects at the organ or cellular level may alter the metabolism of the fish and hence its ability to withstand stress. Even if the fish is not directly affected, any effect on fish-food organisms may result in a starved population of fish. In this chapter such effects of pesticides on fish are discussed.

II. PESTICIDE-INDUCED MORPHO-ANATOMICAL CHANGES

A. Morphological Changes

Many morphological changes have been reported following the exposure of fish either to high concentrations for brief periods, or to sublethal concentrations for extended periods. Some of these changes may be apparently as innocuous as change of the body coloration or may be as gross as erosion of fins, distortion of vision, etc. Small or large, many of these changes can prevent the fish from functioning effectively and efficiently in the environment.

Exposure of a predatory fish, *Therapon jarbua*, to 2 μg/l of DDT for 15 days resulted in the darkening of skin, formation of a brown spot on the forehead, swelling of the eyes, and erosion of the fin margins.[1] DDT-fed juvenile coho and chinook salmon developed severe hyperplasia of the nose. This deformity started as a small whitish spot at the tip of the nose and gradually developed to a stage where one or both the eyes were lost.[2] In the early stages, an acute inflammatory exudate accumulated in the s.c. tissues of the oral surface epithelium and on the tips of both jaws; later this inflammation gradually spread, resulting in surface ulceration. In the final stages, a breakdown of the skin barrier led to a massive invasion of the ulcer by Gram-negative bacteria. Visual observation showed no increased tendency on the part of the fish to bump against the walls of the container, which could have caused such a severe damage to the nose region. Guppies that survived exposure to 0.1 or 10 μg/l TCDD (2,3,7,8-tetrachlorodibenzo-*p*-dioxin, a highly toxic impurity in the herbicide, 2,4,5-T [2,4,5-trichlorophenoxyacetic acid] for 10 days, showed necrosis of maxillary cartilage and fins.[3] Fathead minnows exposed to even one tenth of the 96-h median LC of disulfoton for 24 hr, developed hemorrhaged areas anterior and posterior to the dorsal fin. Rainbow trout exposed to disulfoton were lethargic and developed a darker body coloration.[4] In the same study, Dursban®-exposed trout developed darkened areas anterior and posterior to the dorsal fin and the fish eventually became deformed. Also, fathead minnows exposed to permethrin were found floating with the ventral side up, apparently because of the distension of the peritoneal cavity with gas.

Degeneration of fins and opercles and malformations of the anterior region of the skull were noted in fathead minnows exposed to a commercial composite of penta-chlorophenol (PCP).[5] Shortened head and degenerated fins were also noted. This com-mercial mixture had impurities of tetrachloro dibenzo-*p*-dioxin (TCDD), polychlori-nated dibenzofuran (PCDF), and hexachlorobenzene (HCB), and the observed malformations were consistent with effects reported in mammals exposed to various types of CDDs and CDFs. Young Atlantic salmon exposed to 1 mg/l fenitrothion, swam with distended fins.[6] Bentazon-exposed mosquitofish turned bluish, showed in-tense gill hemorrhaging, and were lethargic.[7] Rainbow trout that survived 96-hr expo-sure to methyl parathion were immobile and had distended abdomens.[8] Swelling of the abdomen of fathead minnows exposed to hexachlorocyclohexane (HCH)[9] and finger-ling rainbow trout exposed to lethal concentration of fenitrothion[10] were reported. In the latter, the distended abdomen contained appreciable amounts of a watery, light yellow liquid in the peritoneal cavity. Methamidophos induced permanently dropped lower jaws in the fingerlings of the common carp.[11]

Fish exposed to phosphate ester hydraulic fluids (pydraul 50 E, pydraul 115 E, or nonylphenyl diphenyl phosphate [NPDPP] and cumylphenyl diphenyl phosphate [CPDPP]) developed cataracts and an opaque eye lens. They also had degenerated retractor muscle of the eye and changed retinal pigmentation. Damage to the retina and the lens and the proliferation of atypical-appearing squamous cells that led to loss of lens function were also noticed.[12] Similarly, anophthalmy, symmetrical and unilat-eral microphthalmy, and curled bodies were observed in the embryos of herring ex-posed to 2,4- or 2,5-DNP (dinitrophenol).[13]

Weis and Weis[14] reported retardation of the regeneration of the caudal fins of *Fun-dulus heteroclitus* following a treatment with 10 μg/l of DDT, malathion, parathion, or carbaryl. The effect of DDT, however, was not as much as that of the others. A relationship between acetylcholinesterase (AChE) inhibition and retardation of the fin regeneration was suggested. Weis and Weis[15] evaluated the relative usefulness of Atlan-tic silverside, sheepshead minnows, and the killifish for monitoring pesticide-induced deformities, and considered the killifish as the most useful species for such a purpose.

B. Vertebral Damage and Anatomical Changes

Many pesticides, irrespective of the group to which they belong, have been reported to induce vertebral damage and skeletal deformities. As early as 1959, Darsie and Cor-riden[16] reported that ocellated malathion-exposed killifish developed a laterally bent caudal peduncle and bent body. McCann and Jasper[17] reported extensive hemorrhag-ing of the vertebral region, accompanied by damage in bluegills exposed to a variety of pesticides. They also reviewed similar reports until 1970. In their experiments, the concentration of six compounds that caused hemorrhage in bluegills ranged from a little over the 24-h LC 50 value in the case of 5% Dyrene® (a triazine) to one fifth of the 24-h LC 50 value in the case of Phosalone®. The earliest time hemorrhaging was detected in bluegills following exposure to Dyrene®, Akton®, Phosalone®, trichlor-fon, demeton, and methyl parathion was 3.75, 12, 6, 12, 6.5, and 4.25 hr, respec-tively.[17] In hemorrhaged areas, the centrum and spinal arches were frequently found fractured, and the surrounding tissue disoriented. A projection of blood cells and con-nective tissue intruded into the lumen of the centrum. Transfer of such fish to recon-stituted water led to the resorption of the hemorrhage within 2 weeks, but bending of the tail section by about 20° continued. Of the six compounds studied, Akton® was the most active hemorrhage-producing chemical and demeton was the least.

Embryos of the sheepshead minnow, exposed to 10 mg/l malathion, developed skel-etal deformities that rendered them incapable of normal swimming. Fry also manifested convulsive, uncoordinated movements, characteristic of the exposed embryos. Such fry, though sometimes normal looking, usually developed a bent appearance akin to scoliosis (lateral curvature of the spine). It was suggested that inhibition of AChE was responsible for this abnormality. Malathion at 1 mg/l and carbaryl or DDT up to 10 mg/l had no such deleterious effect on the developing embryos of sheepshead minnows.[18] Couch et al.[19] reported scoliosis in sheepshead minnows, following exposure to chlordecone, that resulted in severe spinal column injury. All the fish that were exposed to 4 µg/l for 10 days showed signs of scoliosis, loss of equilibrium, and tetanic convulsions. Major spinal column flexure occurred in the vicinity of the vertebrae 17 through 21. Normal muscle bundle patterns were broken in the exposed fish. Inter- and intramuscular hemorrhaging was noticed in severely affected fish. It was suggested that scoliosis was a secondary effect of tetany and paralysis of trunk musculature, under the influence of chlordecone on the central nervous system (CNS). Exposure of sheepshead minnows to as low as 0.78 µg/l chlordecone caused vertebral damage.[20]

Exposure of sheepshead minnows to trifluralin from zygote through the first 28 days of development resulted in vertebral dysplasia.[21] Semisymmetrical hypertrophy of the vertebrae was followed by dorsal vertebral growth into the neural canal, ventral compression of the renal duct, and longitudinal fusion of the vertebrae. Serum calcium levels were elevated in the fish exposed to 16.6 µg/l of trifluralin for 4 days. Vertebral dysplasia was noticed even in fish (*Salmo salar* parr) exposed once to a high level of trifluralin (0.5 mg/l for 11 hr) and maintained in clean water for 12 months. This deformity was comparable to the one seen in fish in nature, following an accidental spillage of trifluralin.[22]

Within 24 to 48 hr after exposure of fathead minnows to chlordecone (10 to 73 µg/l), extensive hemorrhaging, associated with an apparent fracture or dislocation of the vertebral column occurred.[23] Fathead minnows exposed to Dursban® for 48 hr (47 to 383 µg/l) had spinal deformities and stiff vertebral columns. Disulfoton also caused similar deformities in fathead minnows. Spinal deformities in rainbow trout appeared 30 hr after exposure to Dursban®.[4] Vertebral damage and spinal curvature were observed in rainbow trout exposed to triphenyl phosphate for 96 hr.[8]

In a study on the relationship between body constituents and mechanical properties of the bone of striped bass collected from four localities, it was found that vertebrae of those fish that had the highest concentration of contaminants were also the weakest and the most flexible.[24] Such vertebrae had the least elastic strength, lowest rupture point, least stiffness, and least toughness. The body contaminant burden, especially that of polychlorinated biphenyls (PCBs), increased from May to June to the fall period. Vertebrae of fish collected in the spring had significantly greater stiffness, strength, and toughness than the vertebrae of fish collected in the fall. Vertebral damage, resulting from contaminant accumulation, as occurs in nature, seems to lead to impaired swimming and altered feeding behavior, making the organisms more vulnerable to predation.

Pollutant-induced vertebral damage was discussed by Bengtsson.[25] Mehrle et al.[24] suggested that decrease in vertebral mechanical properties is an early indicator of contaminant stress. They also suggested that contaminant-induced competition for vitamin C between collagen metabolism in bone and microsomal mixed function oxidases (involved in the detoxification process of the xenobiotic chemicals) would cause the vertebral damage. The competition for vitamin C would decrease the vitamin C and collagen content of the bone, with a concomitant increase in the ratio of bone minerals to collagen, resulting in an increased fragility of the bone.

III. PESTICIDE-INDUCED BEHAVIORAL CHANGES

A. General Behavioral Changes

Warner et al.[26] commented, "The behaviour (or activities) of an organism represents the final integrated result of a diversity of biochemical and physiological processes. Thus, a single behavioural parameter is generally more comprehensive than a physiological or biochemical parameter." They also considered that, "Behaviour patterns are known to be highly sensitive to changes in the steady state of an organism." Hence, in their search for a rapid, biological method for detecting the sublethal effects of pesticides, they chose alterations in the behavior as the diagnostic tool for identifying the ecological effects of the release of a toxicant into the environment. Scherer[27] considered that behavioral tests may sometimes ascertain lower thresholds than physiological techniques, because the response results from an intact, integrated, functional system.

Bull and McInerney[28] explained the meaning of the various terms like feeding, chewing, coughing, yawning and nipping, chasing, displaying, displacing, vacating, flight, appeasement, flicking, and thrusting. They studied the behavioral changes in coho salmon exposed to several concentrations of fenitrothion (7.3, 18, 36.5, and 57.8% of the 96-h LC 50 value). Various behavioral changes occurred within 2 hr of exposure. A marked decline in the frequency of various agonistic behaviors (chasing, vacating, nipping, etc.) was noticed. On the other hand, comfort behaviors like flicks, thrusts, coughs, etc. increased with increasing concentration of the toxicant, but declined at higher concentrations. Exposed fish showed an altered station selection, with a shift from central position to downstream, near surface, or lateral positions. At higher concentrations, some individuals were unable to maintain position and were swept downstream. After a 5-hr exposure, fish swam near the surface with bloated stomachs and heads downwards. Young Atlantic salmon exposed to 0.1 or 1 mg/l fenitrothion moved less than the controls.[6] Six days after the exposure to 1 mg/l for 15 to 16 hr, 50% of the fish did not hold territories.

Fathead minnows, channel catfish, and bluegills exposed to phosphate esters became hypersensitive to disturbance, fed less, and showed impaired swimming ability.[29] Rainbow trout exposed to TBTO, a molluscicide, showed progressive signs of tiredness and lost positive rheotaxis. Only 10 to 20% of the fish survived upon being transferred to clean water after the fish showed signs of poisoning.[30]

B. Avoidance Reactions

Fish can sometimes sense the presence of a xenobiotic chemical in water and tend to avoid it. In an experiment to test the avoidance reactions of sheepshead minnows to pesticides, the fish avoided DDT, endrin, Dursban®, and 2,4-dichlorophenoxyacetic acid (2,4-D) at three concentrations (one higher and two lower than the 24-h LC 50) but failed to detect the presence of carbaryl or malathion. In the simultaneous presence of two concentrations of the same pesticide in a Y-shaped experimental set-up, the fish avoided the higher concentration of 2,4-D and moved into the lower one, preferred the higher concentration of DDT, and could not detect any difference in the two concentrations of endrin and Dursban®.[31] In a similar test, mosquitofish avoided DDT, Dursban®, malathion, carbaryl, and 2,4-D, but not endrin. This difference between different species of fish in their reaction to different chemicals does not seem to be dictated by the chemistry of the molecule, but appears species-specific. Given the choice of two concentrations of the same pesticide, mosquitofish failed to detect the difference in the concentration.[32]

Sprague and Drury[33] studied the avoidance of four pollutants, viz., alkylbenzene sulfonate (ABS), phenol, chlorine, and kraft pulp mill effluent by rainbow trout and Atlantic salmon. The experiments with chlorine were interesting. Trout avoided 0.01

as well as 1 mg/l; the latter was lethal in 4 hr. Most fish preferred an intermediate lethal concentration of 0.1 mg/l which is a sort of a sensory trap for the fish. Rainbow trout larvae normally avoid light and gather at the darker corners of the containers; after their exposure to Dylox®, they no longer avoided light. This change in behavior, as well as jerky swimming, was observed in all the concentrations of Dylox® tested; the effect was produced faster in the higher concentrations.[34] Likewise, DDT at sublethal oral doses suppressed a reaction brook trout had learned, i.e., to avoid their preferred light or dark sides of the aquarium.[35] Goldfish avoided fenitrothion, at as low as 10 μg/l.[27] Following exposure of goldfish to sublethal levels of parathion, the overall activity of the individuals declined and their swimming behavior and orientation angle were significantly altered.[36] The effect of diquat, simazine, and their formulations at the lowest, highest, and five times the highest field rates was tested on the rheotaxis (orientation in relation to current) and swimming speeds. Diquat and its formulation, Reglone A®, altered the rheotaxis, indicating that sublethal concentrations would cause a downstream drift.[37] The toxicant-induced modification of rheotropism was suggested as a sensitive tool for detecting sublethal effects. Rainbow trout avoided lethal and slightly less than lethal concentrations of another herbicide, glyphosate.[38] When Abate® was applied to River Oti in Ghana to control *Simulium* larvae, fish 300 m upstream from the point of application were normal, whereas fish at the site of application swam erratically. Fewer numbers and species were collected in the 24-hr period following application than before application, suggesting possible avoidance reaction of fish to Abate®.[39]

Rainbow trout fry could detect and avoid copper sulphate, dalapon, 2,4-D (DMA), xylene, and acrolein, but did not show any response to the presence of glyphosate, Aquathol K® (dipotassium salt of endothal), diquat, or trichloroacetic acid (TCA).[40] Since endothall is known to be highly toxic to fish and is one of the commonly used herbicides, failure of fry to avoid endothall can prove disastrous.

Warner et al.[26] described a Conditioned Avoidance Response Apparatus (CARA) and studied the effect of sublethal stress of pesticides. Using a concentration of 1/25 of the 96-h LC 50 of toxaphene or 1/200 of the 96-h LC 50 of tetraethyldiphosphate (TEPP), the behavioral pathology could be demonstrated with this sophisticated apparatus. Methods of quantitative analysis of such behavioral changes were described by the authors.

C. Swimming and Hypersensitivity

In the small streams in Prince Edward Island, Canada, only rarely were the trout and salmon observed together in pools, but subsequent to accidental contamination of Mill River with a mixture of nabam and endrin from a sprayer, many trout of all ages, and salmon fry and parr were observed together in several deep pools. The salmon were found near or at the surface of the pools, whereas normally they live close to the stream bed. Thus, aberrant swimming, gathering in unusual parts of the water column, unusual species associations, unusual response to electric field, and unseasonal downstream movement are some useful biological indicators of pesticide pollution.[41] Exposure of eggs of Atlantic salmon to 0.05 or 0.1 mg/l DDT up to 1 month, resulted in impairment of balance and retarded behavior in alewins. Changed behavioral patterns may have deleterious consequences on the populations, as the altered swim-up behavior and surface orientation may delay emergence which is usually timed to coincide with the period when the yolk sac reserves are almost exhausted. Even if the alewins emerged, they would have had difficulty in feeding and holding position in their natural habitat and would have been extremely vulnerable to predation.[42] Chronic exposure of pinfish to Aroclor® 1216 (42 days to 32 μg/l) induced erratic swimming and sluggish feeding. The affected fish swam near the surface before losing equilibrium and becoming moribund.[43] Adelman et al.[44] considered that the fish that lost equilibrium would either die almost immediately or would lie on the bottom of the aquarium in this condition for a few days and then die. Fish that have lost equilibrium in the environment either fall an easy prey to predators or will be at the mercy of the water currents. The extent of physiological impairment of perch as a result of exposure to a toxicant was tested by recording the ability of the fish to compensate for torque in a rotating current.[45]

Excited swimming and hyperactivity are as deleterious to the fish as lethargic swimming and loss of equilibrium, because the former two make the fish conspicuous in the surroundings. For instance, zebrafish larvae, upon exposure to 1 mg/l captan for 5 min, became strongly excited. After about 30 min of excitement, the larvae became nonmotile and died. Death was always associated with severe head injury in which the eyeballs were blown out of the sockets and the head was ruptured into lateral halves to give a bicephalus appearance.[46] Exposure of brook trout for 24 hr to sublethal doses of DDT (0.1 to 0.3 mg/l) rendered the lateral line hypersensitive to stimulus.[47] Likewise, exposure of goldfish to endrin through food (143 to 430 μg/kg body weight) caused hyperactivity, followed by loss of equilibrium, and hypersensitivity to physical stimuli.[48] Fingerling channel catfish exposed to sublethal doses of endrin were hypersensitive to sudden noise or movement.[49] Guppies exposed to dieldrin showed increased sensitivity to light, movement, and vibration. Their locomotor behavior was characterized by random and rapid swimming.[50] At sublethal concentrations, DDT induced locomotor hyperactivity in the bluegill. Even after the maintenance of the exposed fish in clean water for 2 weeks, this effect was not reversed.[51]

Young Atlantic salmon that were force-fed mealworms contaminated with fenitrothion, did not attempt to avoid capture with a dipnet, whereas it was hard to catch the control fish.[6] Similarly, young shiner perch that developed severe eye defects consequent upon exposure to chlorinated sea water, failed to avoid capture with a dipnet.[52] Following the treatment of the Upper Volta River with Abate®, for the control of *Simulium* larvae, no fish died, but fish in the treated area were prone to easy capture. The catch per unit effort and the number of species captured in the 24-hr period following treatment, were higher. Evidently, sublethal stress leading to under- or hyperactivity, make the fish fall an easy prey in nature and affect the stability of the population.

D. Schooling Behavior

Weis and Weis[53] reported that carbaryl had a marked effect on the schooling behavior of the Atlantic silverside. The group exposed to 100 μg/l carbaryl for 1 or 3 days, consistently occupied twice the area occupied by the control group. This altered schooling behavior persisted until the 3rd day after the transfer of the exposed fish to uncontaminated water. Quoting the earlier work of Zuyev and Belyayev,[54] the authors suggested that a group hydrodynamic effect might exist, which would help the fish to swim within the school. Weis[53] showed that there was a 30% saving of the swimming effort when the fish swam within the school, which diminished the food requirements. Disruption of schooling behavior, due to the sublethal stress of toxicants, results in increased swimming activity and entails increased expenditure of energy, and hence higher food requirements, with attendant serious ecological consequences. After 3 days in 1 μg/l DDT solution, goldfish swam faster than the controls and occupied more space. The experimental fish were more spread out and appeared to be swimming independent of one another. Parallel orientation and even spacing — a characteristic of the controls — was not evident in the experimental fish. Loud sounds also disturbed the experimental fish to a greater extent than the controls. The schooling behavior returned to normal only 7 days after the transfer of the fish to insecticide-free water.

Increased uptake of the toxicant, owing to a higher metabolic rate and higher food consumption resulting from the disruption of the school, would compound the harmful effects of the chemical. This disrupted schooling results in increased predation of the stragglers.

Apart from disruption of the school, pollutants affect the spawning migration of fish.[55] Salmon in unpolluted Tabusintac estuary moved up fairly rapidly compared with their passage through the contaminated Miramichi estuary, in Canada. They turned back when they encountered polluted waters. Such behavior in natural populations can seriously hinder the spawning migration and would severely affect reproductive success.[56]

Erosion of taste buds, apathy to food, and reduced feeding or altered feeding behavior owing to exposure to sublethal concentrations of pesticides are known. These effects are discussed in a later section.

E. Coughing

Coughing in fish has been described as an interruption of the normal ventilating cycle, with a more rapid expansion and contraction of the buccal and opercular cavities, and serves the purpose of clearing the gills of accumulated debris.[57] Coughing frequency in coho salmon increased with increasing concentration of fenitrothion.[28] Carlson[57] considered that this response could also be elicited by mechanical and chemical stimuli, and increased quickly within hours of exposure to a toxicant. The frequency of coughs was proportional to the sublethal toxicant concentration and could predict the chronic effects at levels near the Maximum Acceptable Toxicant Concentration (MATC). In order to understand the fish cough response better and to define the ventilatory characteristics during normal respiration and toxicant stress, Carlson[57] studied the normal pattern of fish cough response in the bluegills with the help of a strip chart recorder. Three types of coughing maneuvers were identified. Further, Carlson and Drummond[58] found fish cough response a useful and sensitive tool for evaluating the quality of industrial and municipal effluents. The coughing frequency of bluegills increased with increasing effluent concentration.[58] Schaumburg et al.[59] noticed a direct relationship between the frequency of coughing in rainbow trout and the time of exposure to DDT. Changes in the respiratory pattern occurred at concentrations lower than those that were lethal in toxicity tests. Variations in the rate of respiration and cough frequency were not interrelated. For instance, at very low concentrations of DDT, the rate of respiration remained unchanged, whereas the cough frequency increased. Drummond et al.[60] could detect changes in brook trout cough frequency at very low sublethal concentrations of copper. Exposure of rainbow trout fingerlings to 0.5 mg/ℓ fenitrothion affected the cough frequency in the early period of exposure (96-h LC 50 of fenitrothion was 2.4 mg/ℓ). In another study, too, fenitrothion increased the coughing frequency of rainbow trout, whereas acephate did not.[61] Likewise, DDT increased the coughing frequency of *Therapon jarbua*.[1] Klaverkamp et al.[10] considered that cough frequency was a more sensitive response than heartbeat and ventilation rates in the adult fish.

F. Learning Ability

DDT produced an increase in amplitude and decrease in the frequency of the spontaneous cerebral activity in the goldfish, resulting in complete loss of balance.[62] Loss of learning ability in brook trout was reported by Anderson and Prins.[63] A weak propeller-like movement of the tail was elicited by administering a mild electric shock to the gular region. Fish were trained to show this response by exposure to light before the application of the shock. Control fish took significantly fewer trials to exhibit the propeller-like tail reflex, whereas a 24-hr exposure to sublethal concentration of DDT resulted in complete impairment of the learning ability of half of the experimental fish. The other half required significantly more trials to exhibit the reflex. Davy et al.[64] observed a significant time-dependent correlation between successive turns in goldfish. This correlation decreased as elapsed time between turns increased. They concluded that this phenomenon indicates the presence of a memory controlled by the CNS. Exposure of goldfish to DDT (10 μg/l) for 4 days significantly reduced this correlation between turns. The correlation was not restored even after 130 to 139 days following the transfer of the fish to clean, uncontaminated water.[64] Hatfield and Johansen[65] studied the effect of four insecticides on the ability of Atlantic salmon parr to learn and retain a simple conditioned reflex. Control parr could be trained to avoid the dark side of a test system, using light as the conditioned stimulus and electric shock as the unconditioned stimulus. Exposure to 96-h median LC of fenitrothion for 24 hr completely inhibited the learning ability of salmon parr; Abate® at 5 mg/l retarded learning: DDT at 0.07 mg/l mildly enhanced learning; methoxychlor had no effect. Orally administered DDT increased the rate of learning of rainbow trout.[66] In a further experiment, rainbow trout were trained to distinguish between normal and dim light. Such trained fish were tested to determine their ability to discern differences in intensity of light, 71 hr after being given an oral dose of DDT. It was observed that the ability decreased with increasing dose. Rainbow trout that were fed a DDT- or DDE-containing diet required significantly less time than controls for recovering from anesthesia.[67]

The effect of pesticides on other behavioral patterns has been reported. Salmon parr exposed to 1 mg/l fenitrothion were more vulnerable to predation by brook trout.[68] Similarly, exposure of the shrimp *Palaemonetes pugio* to methyl or ethyl parathion resulted in increased spontaneous activity of the shrimp and rendered it more susceptible to predation by the killifish;[69] moreover, exposed shrimp were easily fatigued when pursued. Juvenile rainbow trout exposed to > 7 mg/l phenol were more vulnerable to preying by adult rainbow trout than controls or juveniles exposed to < 7 mg/l.[70] Sublethal concentration of phenol had a depressing effect on the courtship behavior of male guppies.[71]

G. Effect of Pesticides on Temperature Selection

Ogilvie and Anderson[72] reported that Atlantic salmon exposed to DDT selected a different temperature than that of the controls. The fish exposed to lower doses selected a consistently lower temperature than that of the controls, whereas those exposed to higher doses ($\geq$ 10 μg/l), selected a temperature higher than that of the controls. This shift appeared to be more marked in the case of fish initially acclimated to higher temperatures (17°C) than those acclimated to lower temperatures (8°C). Fish that were acclimated to higher temperatures and exposed to $\geq$ 10 μg/l DDT became hyperactive when they encountered low temperatures. A similar shift in preferred temperatures was noted with brook trout exposed to DDT.[73]

Brook trout fingerlings normally occupied the extreme ends of a trough when the temperature of water was constant at 10°C. When the temperature gradient was from 5 to 25°C, the fish selected an area with a temperature of 14.5°C. Exposure to 20 to 30 μg/l of *p,p'* and *o,p'*-isomers of DDT and DDD, *p,p'*-methoxychlor, *p,p'*-DDA, and *p,p'*-Cl DDT led to the selection of a lower temperature; *p,p'*-DDE and *o,p'*-DDE had only a small effect on the temperature selected. The effect of prolonged exposure to the toxicants on temperature selection by brook trout was also studied. There was no significant change in the temperature selected by controls in 9 days of observation; 24-hr exposure to 20 μg/l DDT caused a selection of 4 to 4.5°C lower temperature than the normal. With methoxychlor, the initial 5°C lowering of selected temperature was progressively reversed and by the 5th day the selected temperature was not significantly different from that of the controls, which is attributable to the biodegradability of methoxychlor unlike DDT.[74] Peterson[75] noted that in a horizontal temperature gradient, juvenile Atlantic salmon normally chose 14.6 to 17.5°C; upon exposure to DDT, the selected temperature was shifted depending on the exposure concentration. Fish exposed to lower concentrations of DDT either did not choose a temperature different from that of the controls or chose a lower temperature; fish exposed to higher concentrations (20 to 60 μg/ℓ) chose a temperature higher than that of the controls; p,p'-DDD, methoxychlor, o,p'-DDT, and p,p'-DDE also induced a similar change, although the concentration of the compound needed to produce the temperature shift increased in the order indicated. Aldrin, at a concentration of 100 to 150 μg/ℓ lowered the selected temperature. Aroclor® 1254 even at 2 mg/ℓ had no such effect.[75] Malathion induced the common shiner to choose lower temperatures than the controls when the fish were acclimated at 17°C, but not at 8°C.[76] The alteration of the temperature selected by underyearling Atlantic salmon exposed to sublethal concentrations of phenol persisted for at least a month after the transfer of the fish to clean water.[77] Peterson[78] introduced the term Q_{TOX} — temperature quotient of toxicity that is calculated, as follows —

$$Q_{TOX} = \left[\frac{LC\ 50\ (at\ T_2)}{LC\ 50\ (at\ T_1)}\right]^{\frac{10}{T_1 - T_2}}$$

A Q_{TOX} of < 1 indicates greater toxicity at lower temperatures and vice versa. Analyzing the published reports, he concluded that those compounds that increase the preferred temperatures are more toxic at lower temperatures and vice versa. There is one report on the influence of pesticides upon the selection of salinity by fish. DDT-exposed mosquitofish consistently selected higher salinities than unexposed fish. No such change was induced by malathion.[79]

H. Apparatus to Record Behavioral Patterns

A number of devices that could be used for recording changes in the behavioral patterns of fish have been described. The CARA devised by Warner et al.[26] has already been referred to. The fabrication and functioning of this unit as well as the analysis of the data were described in detail. Spoor et al.[80] described a simple electrode chamber which could measure the water currents produced by respiratory and other movements. A simple method to record fish heart and opercular beats, without implanted electrodes, was described by Rommel.[81] Scherer and Nowak[82] devised an apparatus for recording avoidance movements of fish exposed to pollutants. While most of the devices to record avoidance movements of fish require visual observation, Cripe[83] described an automated device for recording avoidance movements. Carlson[57] recorded the electrobranchiograms (EBGs) and bioelectric potentials on a strip chart recorder using noncontact electrodes. Recently, attempts have been made to use the coughing response and respiratory movements of fish as a means of identifying any change in the effluent treatment pattern and consequent influx of untreated and toxic effluent. Cairns et al.[84] described the coupling of a device that can record fish ventilatory and breathing activity with a minicomputer for biological monitoring work. Such automatic monitoring of respiratory and other responses of aquatic organisms is becoming a more common way of assessing toxicant-induced stress.

IV. PESTICIDE-INDUCED BIOCHEMICAL CHANGES

Many pesticides have been reported to produce a number of biochemical changes in fish, both at lethal, and more often, at sublethal levels. Changes in ion concentrations, organic constituents, enzyme activity, endocrinal activity, and osmoregulation in fish have been attributed to pesticides. After nearly 30 years of research on this subject, however, little is known about the ecological significance of such pesticide-induced changes.

A. AChE Inhibition

Of all the pesticide-induced biochemical changes, inhibition of AChE, the enzyme involved in terminating the action of the neurotransmitter acetylcholine, is perhaps the most often studied. This enzyme and its closely related ones are responsible for the toxicity of organophosphate (OP) and carbamate compounds to vertebrates. The action and metabolism of individual groups of compounds was excellently reviewed by O'Brien.[85]

About 75% of the OP compounds are rather poor inhibitors of cholinesterase in vitro, whereas the rest are potent and direct inhibitors. Yet when the actual poisoning of the animals is considered, no such difference is noticeable, because of the in vivo conversion of the poor inhibitors to a more toxic form. This is more often seen in such OP compounds with a P = S linkage (poor inhibitors of cholinesterase) which are activated to the more toxic oxygen analog, P = O.

Quite early in the study of the effects of pesticides on fish, it was noted that the extent of brain AChE reduction was proportional to the concentration of the substance.[86] Weiss,[87] noting that in vivo inhibition of a number of OP compounds was a function of the exposure concentration and duration, suggested that in vivo response of fish brain AChE to low levels of OP compounds could be used to identify the past and present exposure to AChE-inhibiting compounds. It was also suggested that by utilizing suitable species, it may be possible to demonstrate the presence of OP compounds in water at concentrations < 0.1 mg/ℓ.[88] Weiss and Gakstatter[89] investigated the use of inhibition of brain cholinesterase as a bioassay tool to detect low levels of contamination of water with OP compounds. Distressed menhaden collected from nearshore regions of Ashley River, South Carolina, had about 46.8% less AChE activity in brain homogenates than the undistressed ones.[90] Coppage and Matthews[91] considered that instead of depending on the "dubious interpretation of residues alone"; measurement of AChE activity coupled with residue analysis would be helpful in determining the cause of fish kills. In a further study, Coppage[92] reported that inhibition of 87% of the normal activity is necessary to indicate exposure of fish to anti-AChE compounds. Inhibition to $< 17.7\%$ of normal activity resulted in 40 to 60% fish mortality.

Irrespective of the exposure time and exposure concentration, the AChE activity of dying pinfish was reduced by 72 to 79%.[93] Enzyme inhibition was higher in lethal exposures than in sublethal exposures. The observed correlation between brain AChE activity and deaths was considered to be of diagnostic value. After 3.5, 24, 48, or 72-hr exposure to 575, 142, 92, and 58 µg/ℓ of malathion 40 to 60% of the pinfish died and the extent of inhibition of AChE was in the range of 72 to 79%.[93] Methyl parathion caused 58.6 to 83.7% inhibition of AChE within 10 hr after exposure of shiners.[94]

In a further study, Coppage and Matthews[95] noted that at a concentration of naled that caused 40 to 60% mortality of pinfish, the brain AChE inhibition was 84 to 89% of the normal activity. The authors opined that this correlation between the extent of inhibition of AChE and mortality in the population could be used as a diagnostic tool. Further, they felt that the greater extent of inhibition of brain AChE activity (59 to 65%) caused by sublethal exposure facilitated early detection of pollution by anti-AChE compounds. Brain AChE activity was consistently lower in fish caught downstream from the point of entry of the effluent of an OP and carbamate formulating plant when compared with those caught upstream.[96] Dilute effluent induced lower brain AChE activity, and analysis of the effluent showed the presence of AChE-inhibiting compounds. In another study, accumulation of malathion or malaoxon in the tissues of pinfish was negligible, but considerable amounts of malaoxon monoacid accumulated in the liver and gut. A direct relation between the extent of AChE inhibition and the malathion monoacid residue concentration was noticed.[97] It was concluded that routine type residue monitoring of water would not show poisoning by enzyme-bound metabolites, and in such cases determination of AChE inhibition and similar studies alone can reveal the occurrence of hazardous conditions. Holland et al.[98] suggested modifications in the routine AChE assay method and considered it as a good method for periodic surveillance for the presence of AChE-inhibiting compounds in the aquatic environment.

The suggestion of Weiss that brain AChE inhibition could be used as evidence of exposure to anticholinesterase compounds led to considerable controversy. Gibson et al.[99] noted that fish that became moribund in 750 μg/l parathion solution, showed only 25% AChE activation, whereas those that became moribund in 20 μg/l showed 50% inhibition, and concluded that mortality and recovery from OP poisoning were not necessarily related to the degree of AChE inhibition. Hogan[100] showed that the results of bioassay tests using AChE activity were subject to the influence of the ambient water temperature and hence cautioned against relying too much on such tests for monitoring work. An automated pH stat method devoid of the drawbacks of a spectrophotometric method, and which permits the adjustment of pH, temperature, and enzyme and substrate concentration, was described for studying the brain AChE inhibition in fish.[101] One source of error was that although no destruction of the enzyme took place at an alkaline pH, considerable nonenzymatic hydrolysis occurred. Coppage[101] also observed that the in vitro inhibition was not closely correlated with the toxicity of the compound. For instance, diazinon inhibited the AChE activity to the maximum extent among 4 compounds studied, but it was only 1/30 as toxic as guthion. Although inhibition by guthion, phorate, and parathion was in the order of their toxicity, the extent of inhibition of the enzyme was not of the same magnitude as their toxicity.

Benke et al.[102] compared the relative toxicity of ethyl parathion and methyl parathion to mice and fish. The toxicity of both these compounds to mice was almost the same, whereas methyl parathion was much less toxic than parathion to sunfish. The lesser toxicity of the former was due to the differences in the metabolism of the two compounds in sunfish. However, there was no difference in the toxicity of the oxygen analogs (methyl paraoxon and paraoxon) in the sunfish. In another study, among methyl parathion, parathion, and azinophos methyl (APM) the first one produced the most rapid change in AChE activity following an i.p. injection of pumpkinseed sunfish. The maximum level of inhibition was reached within 2 to 4 hr with methyl parathion, at the end of 4 hr with APM, and after 20 hr with ethyl parathion, although the last two compounds are more toxic than the first, to fish.[103]

Inhibition of AChE under experimental conditions was also investigated. Marked inhibition of AChE in rainbow trout larvae exposed to various concentrations of Dylox® was reported.[34] In experimental ponds, following a Dursban® treatment of 0.05 lb/a, the AChE levels of bluegills and largemouth bass were drastically reduced.[104] Chlorpyrifos-induced inhibition of AChE in *Fundulus heteroclitus* was proportional to the exposure concentration.[105] Fenitrothion at various sublethal concentrations (0.75 to 1.78 mg/l) induced 13 to 25% inhibition in two species of fish by the emulsifiable concentrates (EC) formulations of three OP compounds (Rogor®, Zolone®, and malathion 50% EC).[106] The in vitro and in vivo effect of 12 OP and 9 carbamate pesticides on the inhibition of brain AChE activity of topmouth gudgeon was reported.[107] Carbamates had a higher in vitro inhibitory effect than the OP compounds. No correlation between in vivo AChE inhibition and symptoms of poisoning were noted.

Following the exposure of brook trout and rainbow trout to malathion, the AChE levels were approximately 24 and 28% of those of the controls, and the fish had less than one third the ability of the controls to do work. Thus malathion, though not directly and immediately toxic, had a deleterious effect, owing to the impairment of the ability of the fish to maintain equilibrium, to search for food, and to avoid predators.[108]

The cholinesterase activity in the erythrocytes, gills, heart, and serum of rainbow trout was reduced within 3 hr of exposure to acephate and 1 hr after exposure to fenitrothion.[61] Although acephate is a direct inhibitor of AChE, and fenitrothion needs to be activated (an indirect inhibitor), the latter is more toxic than the former, as is evident both from the LC 50 values and also from the extent of inhibition of AChE. It is likely that the greater toxicity of fenitrothion is attributable to its faster and greater uptake than acephate, because of the greater lipid solubility of the former. Also, fenitrooxon, the active inhibitor of AChE, was five orders of magnitude more potent as an inhibitor of in vitro AChE than acephate. With methamidophos, the extent of brain and liver AChE inhibition was proportional to the insecticide concentration and exposure time. Smaller fish started dying when the AChE inhibition was 40 to 50%, but very large fingerlings survived an inhibition of more than 80%.[11] Among the various tissues of *Tilapia*, highest levels of AChE inhibition were noted in brain followed by muscle, gill, and liver, in that order.[109] In nature, subsequent to aerial spraying of acephate to control spruce budworm, no significant depression of brain AChE activity of brook trout and salmon in streams near the target area occurred; however, there was a significant depression of brain AChE activity in suckers, which returned to normal by the 8th day.[110]

Olson and Christensen[111] tested 74 chemicals for their AChE-inhibiting activity in the muscle of fathead minnow. Eserine was the most reactive with AChE; among 13 pesticides studied, carbaryl was the most inhibitive. They concluded that AChE analyses cannot be used unrestricted to monitor OPs and carbamates. For instance, arsenite ion was more reactive than many carbamates. Similarly, a mixture of ten inorganic toxicants had a similar effect as a mixture of ten organic pesticides.

It has already been explained that majority of OP compounds are indirect inhibitors of AChE. One piquant problem in environmental hazard assessment is that there is no reliable method of measuring the actual AChE inhibitors, i.e., the activated forms; the AChE-inhibiting metabolites may persist long after the disappearance of the parent compounds that can be measured by routine residue methods. Such routine methods do not account for the enzyme-bound forms.[94]

B. ATPase Inhibition by Organochlorines (OCs) and Mode of Action of OC Compounds

1. OC-Induced Inhibition of ATPase

Despite nearly 4 decades of research on OCs, and especially DDT, the biochemical basis of their toxic action is not yet unequivocally established. Although early investigations on the action of DDT on ATPase showed no adverse effects, in 1969 Koch[112,113] reported that chlordane, aldrin, dieldrin, dicofol, lindane, and DDT inhibited the ATPase associated with oxidative phosphorylation and cation transport in plasma membranes. Subsequently, it has been well established that DDT acts through its inhibitory effect on ATPases. The inhibition of ATPases by chlorinated hydrocarbons has been recently reviewed.[114]

The plasma membrane ATPases were less sensitive to OCs than the mitochondrial Mg^{2+}-ATPase. At low concentrations, DDT was the most effective inhibitor of the Mg^{2+}-ATPase. At higher concentrations, dicofol produced the greatest inhibition of both Na^+, K^+, and Mg^{2+}-ATPase. Matsumura and colleagues[115,116] also reported similar findings, that DDT inhibited ATPases.

Following the finding that the toxic action of OCs may lie in their inhibition of ATPases, a major controversy developed. Koch and associates[114] reported that DDT was a more effective inhibitor of Mg^{2+}-ATPase than Na^+, K^+-ATPase activity. Employing oligomycin, a specific inhibitor of mitochondrial portion of the total Mg^{2+}-ATPase activity, it was established that DDT completely inhibited oligomycin-sensitive Mg^{2+}-ATPase (OS-Mg^{2+}-ATPase). On the basis of this finding, it was postulated that mitochondrial Mg^{2+}-ATPase was the primary site of action of DDT. On the contrary, Matsumura Ghiosuddin[117,118] suggested that Na^+, K^+-ATPase was the specific site of action of DDT. Subsequently, it was also suggested that DDT had a specific effect on Ca^{2+}-ATPase activity. Both these suggestions have been rejected by Koch and associates. The latter based their conclusions on the studies that showed greater sensitivity of OS Mg^{2+}-ATPase in insects, mammals, and fish. Na^+, K^+-ATPase and oligomycin-insensitive Mg^{2+}-ATPase (OIS-Mg^{2+}-ATPase) were inhibited to a lesser extent by OC pesticides in the tissue preparations of rat, fish, and American cockroach (Table 1). Additional support that mitochondrial Mg^{2+}-ATPase is the site of action of some OC pesticides was adduced by Koch. Antibodies for kelevan, a derivative of chlordecone, were produced by injecting a bovine serum albumin-kelevan derivative into rabbits and chickens. This antibody could be used repeatedly for reactivation of chlordecone-inhibited OS-Mg^{2+}-ATPase in tissue preparations from animals exposed in vivo to DDT. In vivo exposure of fish and other animals also supports the conclusion of Koch and associates that OS-Mg^{2+}-ATPase, in contrast to OIS-MG^{2+}-ATPase and Na^+, K^+-ATPase, is the most sensitive target of OC compounds.

2. In Vitro and In Vivo Studies with Fish

The brain mitochondrial OS-Mg^{2+}-ATPase of bluegill sunfish was almost totally inhibited by DDE and DDT. The inhibition of the nonmitochondrial OIS-Mg^{2+}-ATPase was only one fourth of the inhibition of the mitochondrial Mg^{2+}-ATPase. Aldrin had the least effect on total Mg^{2+}-ATPase but effectively inhibited the nonmitochondrial Mg^{2+}-ATPase.[119] Dicofol, endosulfan, and DDT inhibited the Na^+, K^+-activated Mg^{2+}-dependent ATPase in the heavy microsomal fraction of brain homogenates of rainbow trout. Similar inhibition of ATPase activity by these pesticides was observed in gill and kidney homogenates, too. In the brain, dicofol caused highest inhibition at any concentration; at a concentration of 10^{-5} M, DDT had higher inhibitory effect than endosulfan, whereas at 10^{-4} M the reverse was true.[120] Tetradifon, an acaricide, was a potent inhibitor of mitochondrial Mg^{2+}-ATPase.

In vivo studies on ATPase activity, following chronic exposure of fathead minnows to various concentrations of Aroclor® 1242 and 1254, indicated greater sensitivity of the enzyme at median exposure levels (2.8 $\mu g/\ell$) than at higher (8.3 $\mu g/\ell$) or lower (0.93 $\mu g/\ell$) exposure levels. There was a difference in the in vitro and in vivo response of the enzyme. In vitro, no stimulation of OIS-Mg^{2+}-ATPase activity occurred, whereas in vivo, both OS-Mg^{2+}-ATPase and OIS-Mg^{2+}-ATPase were stimulated in the kidney and liver.[121] Among 14 chlorinated hydrocarbons tested at a concentration of 40 mg/ℓ, chlordane inhibited Na^+, K^+, Mg^{2+}-ATPase of trout gill to the highest extent; 2,4-D and 2,4,5-T did not cause any inhibition. There was a direct relationship between ATPase inhibition and the solubility of OCs in organic solvents. The lipid solubility of these compounds was suggested to be a factor in the magnitude of their inhibition of

Table 1
PERCENT INHIBITION OR STIMULATION OF ATPase BY CHLORINATED HYDROCARBON PESTICIDES[a]

Compound	Source	Tissue	Na⁺,K⁺ ATPase[b]	OS Mg²⁺-ATPase[b]	OIS Mg²⁺-ATPase[b]
DDT	Bluegill	Brain[c]	31	98	50
	American cockroach	Nerve cord	63	100	0
Dicofol	Bluegill	Brain	54	92	60
	American cockroach	Muscle	—	100	90
Methoxychlor	Bluegill	Brain	4	80	—
α-Chlordane	Bluegill	Brain	53	93	69
γ-Chlordane	Bluegill	Brain	61	85	61
Heptachlor	Bluegill	Brain	43	63	40
Telodrin	Bluegill	Brain	68	66	36
Aldrin	Catfish	Brain	42	66	80
Dieldrin	Catfish	Brain	+64	48	48
Photodieldrin	Catfish	Brain	15	61	31
Aldrin transdiol	Catfish	Brain	+5	3	3
Toxaphene	Catfish	Brain	50	59	75
	Bluegill	Brain	44	77	61
	Mouse	Brain	56	55	39
	Mouse	Kidney	42	56	50
	Mouse	Liver	—	42	40
Endosulfan	Bluegill	Brain	32	69	50
Lindane	Bluegill	Brain	14	3	10
Chlordecone	Bluegill	Brain	62	95	60
	Catfish	Brain	73	91	65
	Rat	Brain	100	100	—
	Rat	Heart	100	90	—
	Rat	Liver	—	99	—
Mono alcohol of chlordecone	Catfish	Brain	66	87	54
Mirex	Bluegill	Brain	0	+2	+3
	Fire Ant	Head	+6	12	+3
	Rat	Brain	10	10	—
	Rat	Heart	15	0	—
	Rat	Liver	—	5	—
Mirex derivatives					
10-Monohydro	Fire ant	Head	31	32	54
	Rat	Brain	20	42	—
5,10-Dihydro	Fire ant	Head	33	47	50
	Rat	Brain	28	58	—
8-Monohydro	Fire ant	Head	13	2	32
	Rat	Brain	16	29	—
2,8-Dihydro	Fire ant	Head	16	18	16
	Rat	Brain	28	41	—
Tetradifon	Bluegill	Brain	9	100	—
Oligomycin	Bluegill	Brain	56	100	—
Plictran	Bluegill	Brain	92	100	+31
	Spider mite		97	100	+60

[a] Tested in vitro at 20 μM conc of the pesticide on ATPase activities from different tissues.

[b] A plus sign preceding a numeral indicates stimulation by chlorinated hydrocarbon pesticides.

[c] Results of toxaphene on mouse tissue based on 15 μM rather than 20 μM conc.

From Cutkomp, L. C., Koch, R. B., and Desaiah, D., in *Insecticide Mode of Action*, Academic Press, New York, 1982. With Permission.

ATPase. The in vivo toxicity of the insecticides, however, had no direct relation to the extent of inhibition caused. For instance, dieldrin and endrin, which are very toxic to fish, were very poor ATPase inhibitors. Another significant and interesting observation was the lack of ATPase inhibition when DDT was administered orally, although many test fish died within 36 hr, and tissue residue concentration was as high as 1 to 2 mg/kg.[122] Contrary to the findings of Koch and associates, Campbell et al.[123] reported that DDT significantly inhibited Na^+, K^+-activated ATPase in the kidney and gills of rainbow trout. Desaiah and Koch[124] observed that different types of OC compounds produced different types of inhibitory effect on ATPases; chlordane and dicofol inhibited both Mg^{2+} and Na^+, K^+-ATPases; DDT inhibited mostly OS-Mg^{2+}-ATPase; endrin and isodrin had little effect on Mg^{2+}-ATPase or Na^+, K^+-ATPase in fish brain. In vitro, toxaphene inhibited OIS-Mg^{2+}-ATPase of catfish brain to a greater extent than OS-Mg^{2+}-ATPase. Toxaphene had less inhibitory effect on kidney than on brain ATPases. Kidney mitochondrial Mg^{2+}-ATPase activity was stimulated at low concentrations of toxaphene, but inhibited at higher concentrations.[124] In another investigation, maximum extent of inhibition of catfish brain ATPase activity was caused by aldrin among the cyclodienes studied. The effect of aldrin was more on Mg^{2+}-ATPase than on Na^+, K^+-ATPase. Depending on their concentration, photodieldrin and dieldrin showed different degrees of inhibitory effect on Mg^{2+}- or Na^+, K^+-ATPase. Aldrin transdiol, a metabolite of dieldrin, did not inhibit any of the ATPases.[125] With chlordecone, Mg^{2+}-ATPase had the greatest sensitivity, and the extent of inhibition was proportional to the concentration of the insecticide.[126] The in vitro inhibition of OS-Mg^{2+}-ATPase was highest with chlordane, heptachlor, and DDT with brain homogenates of bluegills.[27] The synthetic pyrethroids, resmethrin and dimethrin, also inhibited OS-Mg^{2+}-ATPase of bluegill brains to a greater extent than Na^+, K^+-ATPase and OIS-Mg^{2+}-ATPase was not affected. A third synthetic pyrethroid, bioresemethrin, was least effective. With natural pyrethrum, Na^+, K^+-ATPase was more strongly inhibited. Here again, the authors found it difficult to explain the lack of correlation between toxicity of a compound and the extent of inhibition of ATPase.[128]

Many of the above studies were performed on tissue homogenates, and it is possible that the observed effects may be diferent form in vivo ones. Desaiah et al.[129] exposed 45-day old fathead minnows to two different concentrations of DDT in water or food. Brain ATPase analyses made on days 56, 118, 225, and 266 confirmed the in vitro studies and showed that the brain OS-Mg^{2+}-ATPase was inhibited to a greater extent than OIS-Mg^{2+}-ATPase or Na^+, K^+-ATPase. In gills, however, all three ATPase were inhibited, with the mitochondrial Mg^{2+}-ATPase showing the greatest decline. In vivo, the inhibition of all three ATPases was similar in *Labeo rohita* and *Saccobranchus* sp. exposed to chlordane.[130] Jackson and Gardner[131] pointed out that though the OCs inhibited the ATPase activity, they did so at concentrations that far exceeded their solubility limits. In the presence of corexit 7664, a surfactant which completely solubilized DDT, DDE, or DDD, no inhibition of the enzyme was noted. In the absence of the surfactant, the compounds caused a marked decrease in the ATPase activity of in vitro salmon tissue preparations. Methoxychlor had very low inhibitory effect. On the other hand, DDE, which was not highly toxic to fish, had a high in vitro inhibitory effect. Jackson and Gardner[131] wondered, in view of the above inconsistent results, whether the inhibitory effect was due to the precipitation of the enzyme, as indicated by the high turbidity of the enzyme assay mixtures.

While the criticism of Jackson and Gardner[131] that Koch and associates used concentrations of DDT far above its water solubility limits may be justified, surprisingly, in a later study, Jackson and Gardner[132] themselves used 75 μM DDT concentration to study its effect on trout brain Mg^{2+}-ATPase. In such a case, as pointed out by Cutkomp et al.,[114] the enzyme would have precipitated in their system, too. Jackson and Gardner[133] investigated the temperature dependence of trout brain mitochondrial Mg^{2+}-stimulated ATPase. DDT and its analogs caused an increase in the energy of activation and frequency factor of the enzyme-catalyzed reaction, which leads to a negative temperature coefficient of inhibition.

Desaiah and Koch[134] reported an interesting phenomenon, viz., the inhibition of ATPase by DDT and other OCs when certain organic solvents were used to introduce the toxicant into the enzyme assay mixture, whereas some other solvents seemed to reverse the inhibitory effect of the toxicant. For instance, when toxaphene (20 μM) or DDT (10 μM) was introduced through cyclohexane, the inhibition of catfish brain ATPase was 12.5 and 4.1%, respectively, whereas with ethanol it was 53.5 and 47.6%, respectively. Similar observations were made with preparations of cockroach nerve cord and muscle. Dalela et al.[135] also reported that the inhibitory effect of chlordane was greatly reduced [sic] by cyclohexane, cyclopentane, and benzene and to a very little extent by acetone:ethanol (1:1). No such effect of the solvent was observed with OP compounds. Dalela et al. stated, "From the observations made it can be inferred that solvent effect is attributable more to organochlorine pesticides as compared to organophosphates and other classes of pesticides." Desaiah and Koch stated that the type of solvent used can have a great influence on the inhibitory action of pesticides on ATPases. In both these instances, the explanation for the peculiar reversal of pesticide-induced ATPase inhibition, in the presence of some solvents, is quite simple and is not a characteristic of the class of pesticides, as opined by Dalela et al. When 5 to 10 μM of DDT was introduced into a 3- to 5-mℓ enzyme assay mixture along with 1 to 10 $\mu\ell$ of solvent, it amounted to introducing 5.3 to 10.6 μg of DDT along with 1 to 10 $\mu\ell$ of solvent. In terms of concentrations employed in aquatic toxicity tests, this is equivalent to the presence of 1.8 to 3.6 mg of DDT/ℓ in aqueous medium along with 330 to 3330 $\mu\ell$ of solvent per liter. These amounts may be compared with the aqueous solubility of DDT, i.e., 1.2 μg/ℓ. In the presence of a highly nonpolar solvent like cyclohexane or cyclopentane, a nonpolar compound like DDT readily partitions into the solvent phase, and as a consequence, has little contact with the enzyme molecules in the aqueous phase and hence inhibits the enzyme least. On the other hand, in the presence of water-miscible solvents like acetone or ethanol, the pesticide molecule is in close contact with the enzyme molecule and hence inhibits the latter to the maximum extent possible. That the poor water solubility of DDT and other OC compounds is the reason for their noninhibition at ATPase in the presence of nonpolar solvents, and not the reversal of the inhibitory effect of the pesticide by the solvent, is further confirmed by the fact that no such apparent inhibitory effect of the solvent was observed by Dalela et al.[135] in the case of the more water-soluble OP compounds.

The above observation of Desaiah and Koch, and Dalela et al. focuses attention on a question originally raised by Jackson and Gardner,[131] i.e., whether the compounds are truly soluble in the enzyme assay mixture at the amounts employed or whether they exist as colloidal suspensions, micelles, or are adsorbed. For instance, in the case of DDT, as against the known water solubility of 1.2 μg/ℓ, there is no way of retaining 5 μM or 5.3 μg of DDT in true solution, in a 3-mℓ assay mixture. It was recently shown that acetone, at a concentration recommended for aquatic toxicity tests ($\leqslant$ 500 $\mu\ell/\ell$), does not enhance the water solubility of hydrophobic compounds.[136] Hylin[137] emphasized that a compound that is not in true solution is not truly bioavailable. Besides, the likely interaction between the toxicant and the higher concentration of the carrier solvent has to be considered. Although in all the pesticide-induced ATPase inhibition studies, it was categorically stated that the carrier solvent had no adverse effect on the enzyme activity, the possibility of alteration of membrane permeability at the highest concentrations of the carrier solvent employed (a maximum of 3300 as against 500 $\mu\ell/\ell$ recommended for conducting aquatic toxicity tests[138]), cannot be ruled out. These two aspects, viz., the extent of solubility of the OC compounds in a 3- to 5-mℓ enzyme assay mixture and the possible interaction between the solvent and the toxicant (more-than-additive effect), and the alteration of membrane permeability, need to be investigated thoroughly.

Finally, a word of caution in interpreting the results of ATPase studies may be in order. Ewing et al.[139] reported that the gill Na$^+$, K$^+$-ATPase activity in young chinook salmon could be altered by changing the photoperiod at which the fish were reared. Altered growth rates, too, affected the gill ATPase activity. Advanced photoperiod and temperature affected the gill Na$^+$, K$^+$-ATPase in the juvenile steelheads.[140] The ATPase activity of juvenile chinook salmon was affected by feeding.[141] Fish that were fed larger rations showed a peak activity in October and July, whereas no such peak existed in fish that were fed smaller rations. Thus, in any experiments on pesticide-induced changes in ATPase activity, rearing conditions, rearing density, feeding rate, seasonal effect, etc. have to be taken into consideration.

To summarize, the studies of Koch, Cutkomp, and Desaiah have clearly established that ATPase, especially the Mg^{2+}-activated mitochondrial ATPases, are the target sites for the action of many OCs. Although Na$^+$, K$^+$-ATPases and also Ca^{2+}-ATPases were suggested to be the target sites, supporting evidence to that effect is meager. What was thought to be the influence of the solvent in reversing the inhibitory effect of certain OC compounds, seems attributable to the poor water solubility of the compound and the consequent partitioning into the nonpolar phase. The effect of the toxicant on the enzyme at levels that are soluble in the enzyme assay mixture needs to be investigated, although it has been clearly established that OC compounds do inhibit the ATPases.

C. Studies with Other Enzymes

Although the different classes of pesticides have different target sites of action through which they manifest their toxic effect, many pesticides are reportedly metabolic stressors, too. Hence, they have a secondary effect on the enzymes of glycolytic pathway, acid and alkaline phosphotases, enzymes that mediate the tricarboxylic acid cycle, electron transport, and those that are involved in nitrogen metabolism or in innumerable other metabolic pathways.

Endrin altered the glucose-6-phosphatase, alkaline and acid phosphatase, amylase, lipase, and succinic (SDH), pyruvic (PDH), and lactic dehydrogenase (LDH) activity in different tissues of *Channa punctata*.[142-146] Pesticide-induced enzymatic changes were also reported in (compound in parentheses) sailfin mollies (dieldrin);[147] fingerling catfish (Aroclor® 1254);[148] *Tilapia* (methyl parathion);[149-151] *Saccobranchus* (chlordane and metasystox);[152] channel catfish (mirex);[153] *Clarias* (malathion);[154] mosquitofish (malathion);[155] *Channa* (phosphamidon,[156] quinolphos,[157] and diazinon[158,159]); carp (malathion formulation);[160] and *Notopterus* (PCP and phenol).[161-163]

It is difficult to understand the significance of the reported alterations of the enzyme activity, for the enzyme inhibition was rarely reported to be concentration-dependent. Difference in the in vitro and in vivo effects have often been reported. For instance, exposure of yellow eel for 4 days to 0.1 mg/ℓ PCP, decreased the activities of pyruvate kinase and LDH and increased the activities of hexokinase, glucose-6-phosphate dehydrogenase (G-6-PDH), 6-phosphogluconate dehydrogenase, fumerase, and cytochrome oxidase; in vitro studies showed, however, inhibition of activity of all the above-mentioned enzymes, indicating that in vitro studies on enzyme activity do not necessarily reveal the pesticide-induced stress under natural conditions.[164] After studying the inhibitory effect of 13 different chemicals on G-6-PDH, Schultz and Harman[165] concluded that although the test is quite simple, the in vitro inhibition of G-6-PDH cannot be used to predict the toxicity of chemicals to aquatic fauna. While analyzing the effect of pesticides on enzymes in vitro, another point, viz., that the compounds were sometimes used in amounts in excess of their solubility limits, has to be borne in mind. It was once claimed that DDT inhibition of carbonic anhydrase (CA) was responsible for egg shell thinning of raptorial birds. Later, Pocker et al.[166] conclusively demonstrated that DDT, DDE, and dieldrin were not true inhibitors of CA action, but these pesticides, when added in excess of their solubility limit, occluded and precipitated the enzyme from solution. Likewise, Meany and Pocker[167] showed that inactivation of beef, rabbit, and salmon isozymes of LDH occurred only when amounts in excess of the solubility of dieldrin, chlordecone, mirex, *o,p'*- DDT, and *p,p'*-DDT were added. Addition of a surfactant or ethanol as a carrier solvent helped in partial or complete solubilization of the compound, and decreased or eliminated the observed inactivation of LDH. They concluded that the inactivation of LDH by the various OC compounds in aqueous media arose from enzyme coprecipitation and that the observed enzyme inactivation was not a chemical inhibition. They cautioned, "In conclusion, when the effects of compounds such as DDT, DDE and dieldrin, where a very low solubility in water is encountered, are being studied one must be alert to a number of biochemical and physiological consequences that arise from a purely physical action of precipitation and occlusion."

D. Pesticide-Induced Changes in Biochemical Factors
1. Changes in Body Lipid, Protein, and Glycogen Content

Influence of pesticides on a number of biochemical factors has been studied, but the results are rather confusing and contradictory. Prolonged feeding of juvenile coho and chinook salmon (up to 95 days) with DDT increased the lipid levels of the carcass.[2] Yearling coho salmon that were fed PCBs had significantly increased liver lipid reserves, but lowered carcass lipid levels relative to those of the controls. In coho that were fed mirex, liver lipid reserves were unaltered, but carcass lipid was markedly lower.[168] In *Tilapia* exposed to methyl parathion, total lipids and phospholipids decreased, whereas free fatty acids and total cholesterol showed elevated levels in red muscle, gill, liver, and brain.[169] In a short-term exposure of *Channa punctata* to technical-grade endosulfan and its isomers, the former decreased the liver lipid content; isomer-A produced more drastic changes than isomer-B.[170]

Decrease of the protein levels of four tissues of *Channa* sp. following acute exposure to technical-grade endosulfan was observed.[170] Sublethal concentrations of dieldrin did not affect the protein elaboration of rainbow trout.[171] Orally administered DDT caused elevated serum amino acid levels.[172] Liver phenylalanine hydroxylase activity was decreased, and serum phenylalanine concentration was increased in rainbow trout exposed to dieldrin through diet.[173] Exposure to malathion caused little alteration of hepatic protein in *Clarias batrachus,* but there was a marked increase in the free amino acid level, with the incorporation of lysine into the protein of liver being drastically reduced.[174] A decrease in total protein and increase in free amino acid levels in *Channa punctata* exposed to phenthoate were observed.[175] Increased levels of free amino acids were reported in malathion-exposed *Tilapia,* too.[176]

Under the influence of 2,4-D, herring embryos metabolized more carbohydrates in 1 day than they would normally do in the whole embryonic stage.[177] Marked decrease in liver glycogen and serum glucose followed sublethal exposure of an air-breathing catfish to malathion.[174] Technical-grade endosulfan, at lethal concentrations, significantly decreased the liver and muscle glycogen, but increased that of the kidney of *Channa punctata.*[170] Decreased muscle glycogen caused by endosulfan was observed in a catfish, too.[178] On the contrary, phosphamidon increased the liver and muscle glycogen content of *Channa punctata.*[157]

Adult rainbow trout that were given different doses of endrin for 163 days and then forced to swim for 1 hr, had increased liver glycogen; several other biochemical factors were also affected. But it is difficult to attribute the changes only to pesticide stress, because forced swimming itself altered 9 out of 16 serum factors analyzed. Mobilization of liver glycogen was inhibited in endrin-fed trout.[179] Muscle glycogen of a catfish exposed to methyl parathion decreased; so was the case initially with liver glycogen, but later there was a resynthesis of hepatic glycogen.[180] Decrease in the carbohydrates and glycogen of several tissues of *Tilapia* exposed to methyl parathion was also reported.[148]

2. Changes in Ionic Balance

Adult northern puffers exposed to sublethal concentrations of endrin for 96 hr had elevated serum potassium, sodium, calcium, magnesium, and zinc levels concomitant with lowered concentrations of sodium, potassium, calcium, magnesium, and zinc in the livers. These results indicate the mobilization of the cations from the liver into the serum.[181] Altered ionic balance was reported in northern puffers exposed to methyl parathion.[182] In mirex- and PCB-fed salmon, elevated Mg^{2+} concentration, but not altered Ca^{2+} level, was observed.[183] Differences in the Ca^{2+} and Mg^{2+} concentrations of coho salmon between fish from different lakes of the Great Lakes, and also seasonal difference related to spawning, make the interpretation of pesticide-induced changes difficult. Dietary mirex (150 mg/kg dry feed) or PCBs (500 mg/kg dry feed) increased the body water content of rainbow trout, but not that of salmon.[183] Malathion-exposed *Tilapia* had decreased body sodium, potassium, and calcium levels.[184] Similar results were noticed in methyl parathion-treated *Tilapia*.[185]

Exposure of *Sarotherodon* sp. to sublethal concentrations of benthiocarb increased the ammonia, urea, and glutamine levels in all the tissues.[186] On the other hand, ammonia and urea levels decreased in the tissues of the same species exposed to malathion[187] and methyl parathion.[188]

3. Other Changes

Chromosomal aberrations were noticed in mudminnows exposed to the water of the Rhine River, containing several pollutants.[189] Altered DNA and RNA content of muscle, liver, and gill of *Tilapia* exposed to sublethal concentrations of malathion, was noted.[187] Disruption of nucleic acid and protein metabolism occurred as a result of 96-hr exposure to sublethal concentrations of several toxicants. After a detailed analysis of RNA, DNA, and protein content of pollutant-exposed fathead minnow larvae, Barron and Adelman[190] concluded that RNA/DNA ratio is a sensitive index of toxicant stress.

Chronic exposure of goldfish to sublethal levels of endrin caused osmoregulatory failure.[48] DDT impaired the fluid absorption in the intestinal sacs of eels adapted to sea water. DDT also inhibited Na^+, K^+-ATPases and elevated MG^{2+}-ATPase. Disruption of osmoregulation was suggested to be one of the important means by which DDT affects marine fish.[191] Freshwater-acclimated rainbow trout showed no impaired osmoregulation when exposed to 0.33 or 1 mg/ℓ DDT; those acclimated to sea water and exposed to 0.33 mg/ℓ DDT also were normal, whereas fish maintained in 33% sea water and exposed to 1 mg/ℓ DDT showed a significant increase in osmolality over the controls. A similar effect was evident in fish maintained in 100% sea water and exposed to the lower concentration.[192] Intraperitoneally injected DDT caused increased plasma osmotic concentration and death in black surfperch, but only at concentrations higher than the naturally occurring ones. Death as a result of the failure of osmoregulation following exposure to DDT was discounted.[193] Neufeld and Pritchard[194] observed that DDT did not affect Na^+, K^+-ATPases in vivo, but did so in vitro in the blue crab; but in the rock crab, Na^+,K^+-ATPase was inhibited under both in vitro and in vivo exposures.

Mehrle, Mayer, and others[195-200] investigated the effect of sublethal exposure of fish to toxaphene and other contaminants on collagen metabolism and bone composition. Collagen content of the backbone of fathead minnows was reduced following chronic exposure to very low concentrations of toxaphene (55 to 1230 ng/ℓ), leading to a weak backbone.[196] Eyed eggs of brook trout exposed to toxaphene until hatching, hatched into fry with decreased collagen content. The ratio of mineral to collagen increased, rendering the vertebral column brittle.[197] Bone collagen content was reduced in several species of fish by Aroclor® 1254, dimethylamine salt of 2,4-D, and di-2-ethylhexyl phthalate, too.[199] Further, toxaphene reduced the vitamin C content of vertebrae in channel catfish, but did not affect that of liver.[199]

E. Effect on Endocrine Functioning

Moccia et al.[201] found very high frequencies of goiter in coho salmon collected from Lake Ontario and Lake Erie, but could not attribute it to low iodine availability in the respective lakes. Coho from Lake Michigan had a goiter frequency of 6.3%, and the iodine concentration in water was the lowest among the three lakes. Lake Ontario and Lake Erie waters had an iodine content of 2.9 and 1.7 μg/ℓ, respectively, and the goiter frequency of coho from these lakes was 47.6 and 79.5%, respectively. They suggested that the presence of environmental goitrogens, possibly pollutants, may be involved in the production of goiter. Coho salmon, upon being fed 50 μg mirex per gram dry feed, for 2 or 3 months, had a remarkably reduced serum triiodothyroxine (T_3) and serum thyroxine (T_4) levels. Serum T_3 levels were also reduced in PCB-fed fish (500 μg/g feed). Fish that were fed a combination of 5 μg mirex + 50 μg PCB also had reduced T_3 and T_4 levels.[202] In another study, no effect of PCB on thyroid histology was seen, but mirex affected serum T_4 and T_3 levels, without an accompanying change of the histology of the thyroid.[203] PCBs at 50 μg/g had no effect, whereas at 500 μg/g dry feed, serum T_4 and T_3 levels were reduced.[204] Adult coho salmon, fed upon a diet of Great Lakes coho salmon, had evidence of changes in thyroid physiology.[205] Rainbow trout that were given a diet of Great Lakes salmon showed no signs of thyroid hyperplasia nor changes in serum T_3 or T_4 levels.[206] It is not clear whether increased body burdens of OC contaminants cause increased frequency of goiter in nature, because tissue burdens of nine OC compounds measured in fish collected from different lakes did not correlate with goiter frequency.[207] In order to assess the possible risks to human health, coho salmon collected from the Great Lakes were fed to rats and their thyroid response was studied. In female rats there was no adverse effect on the thyroid, but in males fed with coho from any of the three Great Lakes (Lake Michigan, Lake Erie, and Lake Ontario), the thyroid weight was significantly higher. However, there was no significant difference in serum T_3 levels or serum T_3/T_4 ratio.[208]

Freeman and Sangalang[209] proposed that the alteration of steroidogenesis in fish can be used as an indicator of sublethal stress of toxicants, by comparing the steroid hormone metabolism of the fish under investigation with that of normal fish of comparable size, sex, state of maturity, and the like. They described methods to study the steroid hormone metabolism in fish under normal and stress conditions. The steroidogenesis in the testes of brook trout was altered by exposure to 0.2 mg/ℓ Aroclor® 1254.[210] When Atlantic cod were fed for 5 1/2 months (3 times a week) with 10, 50, 100, 250, or 500 μg Aroclor® 1254, the biosynthetic patterns of steroid hormones were altered.[211] Freeman et al.[212] considered that long-term exposure to low levels of pollutants can alter the normal functioning of an organism and destroy the population as effectively as a lethal dose.

Endrin depressed the thyroid activity in goldfish and decreased the thyroid cell height.[48] Aroclor® 1254 altered the thyroid activity in channel catfish.[213] Exposure to one half of the 96-h median LCs of endrin and malathion formulations retarded the gonadotrophin secretion in *Heteropneustes fossilis* and this led to a reduced ovarian ^{32}P uptake.[214] The fish had reduced thyroid activity. After 4 weeks of exposure to a concentration of endrin or malathion that had no effect in 96 hr, the thyroid ^{131}I uptake and the conversion ratio (CR) of protein bound ^{131}I in blood serum, in relation to total serum ^{131}I, were significantly reduced. At half the 96-h median LC, both compounds reduced the pituitary and serum TSH (thyroid-stimulating hormone) content. The 96-h no-effect concentration of a malathion formulation had no effect on pituitary TSH, but reduced the serum TSH level. A 96-h no-effect level of endrin formulation reduced both TSH levels.[215] After 4 weeks of exposure to half the 96-h median LC of endrin and malathion formulations, ovarian ^{32}P uptake was reduced in the preparatory, pre-spawning, and spawning phases. During late postspawning phase, only endrin reduced the ovarian activity. These two pesticides seem to interfere with gonadotrophin secretion.[216] The 96-h no-effect concentration of aldrin and parathion reduced ovarian ^{32}P uptake and total gonadotrophin in the pituitary gland and the blood serum of *Heteropneustes fossilis*.[217] In the males of this species, a 96-h no-effect level of endrin or parathion reduced the testicular ^{32}P uptake and the gonadotrophic potency of the pituitary gland. Exposure to the pesticides elevated the testicular lipid and also the cholesterol content of liver, testes, and blood serum.[218] Endrin and parathion had no effect on the ovarian lipid during the four reproductive phases, i.e., preparatory, prespawning, spawning, and postspawning. During prespawning and spawning stages, ovarian cholesterol was elevated while liver cholesterol was unaffected. Serum cholesterol was elevated during preparatory and late postspawning stages.[219] Formulations of malathion, parathion, endrin, and aldrin caused a significant decrease in ovarian ^{32}P uptake, total gonadotrophic potency of the pituitary gland, and blood serum and the gonadotrophin-releasing hormone-like activity of hypothalamus.[220]

To summarize, pesticide-induced biochemical changes have proved to be a fertile area for research. The ecological significance of the observed changes are, however, difficult to assess. Not only does a compound produce a different response in another organism as revealed by the foregoing discussion, but often with the same species, elevation at lower levels of exposure, inhibition at higher levels, or vice versa is often noted. Furthermore, it is difficult to accept that a deviation from normal is always detrimental. In the absence of adequate information about the normal range of expression of a character or function, it is difficult to appreciate the ecological significance of the observed biochemical changes. Until the natural variation of the biological and physiological parameters in fish is understood, one cannot relate the pesticide-induced biochemical changes to a meaningful assessment of the impact of pesticides on populations. Another justified criticism often voiced against many experimental studies on pesticide-induced biochemical changes is that the levels of pesticides tested for their effect have no relevance to the concentrations occurring in the environment.

V. PESTICIDE-INDUCED HEMATOLOGICAL CHANGES

In human clinical studies, standard methods have been developed and mean values are known for the various biochemical and hematological factors for both sexes, at different ages, and under various conditions of stress. The usefulness of such clinical diagnostic information in the treatment of poisoning cases is well known. Until now it has not been possible to assess the relative well-being of fish and use it as an indicator of the relative well-being of the aquatic environment. If adequate information is available on the various physicochemical and biochemical parameters of fish blood, it may be possible to develop comparable clinical diagnostic tools for fish, too. Christensen et al.[221] used brook trout to evaluate, refine, or develop biochemical procedures for the analysis of fish blood. Analyses were made on leukocytes, erythrocytes, blood protein electrophoretic patterns, erythrocyte sedimentation rate, erythrocytic osmotic fragility, density, surface tension, hemoglobin content, and packed cell volume of blood.

Determination of blood residue levels as a means of identifying previous exposure to pesticides was suggested.[222-224] Witt et al.[225] found a good correlation between DDT levels in blood and those in the adipose tissue. Bridges et al.[226] found 2.7 mg/l of total DDT in the blood of black bullhead, months after the treatment of a pond with DDT.

Upon entering the circulatory system, pesticides rapidly bind to the blood proteins.[227,228] High affinity of methyl mercury for red blood cells was reported.[229] The hematocrit values in bluegills from ponds treated with 10 mg/l 2,4-D were above normal for 3 days, but returned to normal 7 days after spraying.[230] In PCP-exposed eels, both hematocrit and hemoglobin values were higher than in controls. The higher values persisted even after the fish were returned to uncontaminated water. Plasma inorganic phosphate and protein content were also significantly altered. Plasma cholesterol, triglycerides, and free fatty acid levels were elevated in PCP-exposed fish. Hyperglycemia and increased blood glucose was evident, which persisted for 55 days after transfer to clean water.[231] These changes might as well have been caused by the highly toxic impurities (chlorinated dioxins) present in the test material. When eels were exposed to a high dose of PCBs at 1000 mg/kg body weight on alternate days for 1 or 2 weeks, hematocrit values decreased and mean corpuscle hemoglobin concentration increased. On the other hand, microhematocrit values were unaffected in DDT-exposed bluegills.[232] Eels treated at a lesser dose (200 mg/kg body weight) did not manifest any such change.[233] Rainbow trout exposed to bis(tributyltin oxide) (TBTO), an organo-tin compound, showed an increase in packed cell volume accompanied by an increase in the hemoglobin content.[30] A sublethal concentration of malathion induced decreased erythrocyte and increased leukocyte count, and increased serum free amino acid and glucose levels in a freshwater catfish.[174] A 1-hr exposure of sockeye salmon smolts to 5 mg/l of butoxy ethanol ester of 2,4-D did not produce any change in the hematological factors. In a second experiment with slightly larger fish, plasma cortisol increased markedly.[234] *Channa punctata* exposed to sublethal levels of quinalphos for 15 days had lowered hemoglobin content.[157] After 30 days blood glucose, lactic acid, and hemoglobin content decreased. Diazinon exposed *Channa* species showed increased hematocrit values, hemoglobin content, blood glucose, and urea. After 15 or 30 days exposure, blood cholesterol levels also increased.[235,236]

Rainbow trout exposed to the insect growth regulator, diflubenzuron, had reduced serum glutamate oxaloacetate transaminase which was dose-dependent. The serum glucose content of fish exposed to methoprene was also significantly reduced.[237] Rainbow trout, given a single oral dose of PCB (1000 mg/kg), had decreased δ-amino-levulinic acid dehydratase activity after 30 days. Hematocrit and hemoglobin content were unaffected.[238] Pesticide-induced serum ionic changes were also reported in the flounder.[23]

Pesticide-induced hematological changes, such as those described above, may be of some value in assessing the impact of exposure under natural conditions and may also serve as tools for biological monitoring. Uthe et al.[240] stated, "The appearance in blood of such cellular enzymes that are normally bound to cellular structures, indicates more severe cellular damage than the appearance of cytoplasmic or soluble enzymes." Handling stress caused significant changes in plasma glucose Ca^{2+}, Cl^- and cholesterol levels[241] in all the hematological parameters studied,[242] and resulted in body weight changes.[243] Stress due to anesthesia was also shown to produce such changes.[244] Hence, caution should be exercised in interpreting the data on hematological changes. In a study on methyl mercury-induced hematological changes in rainbow trout, a marked decrease in plasma K^+ in the controls in the 4th and 8th week of the experiment and a significant increase in plasma Mg^{2+} concentration in the 12th week were noted.[245]

Against such variations in the controls, it is difficult to assess the impact and significance of any observed change. Weil and Carpenter[246] showed that statistically significant differences between controls and treated animals were attributable to abnormal values in the former. They suggested that to be indicative of deleterious effect, changes produced in treated animals should be dose-dependent, and illustrate a trend away from the norm of the stock of the animals. Hence, it has to be emphasized that data on hematological changes have to be interpreted carefully.

VI. PESTICIDE-INDUCED HISTOPATHOLOGICAL CHANGES

Eller[247] described endrin-induced histopathological changes in cutthroat trout and reviewed the gill lesions in freshwater teleosts. Couch[248] reviewed the histopathological effects of pesticides and related chemicals on the livers of fish and concluded that many of them were nonspecific. No characteristic trends of pathological changes for any class of pesticides were seen. On the other hand, examining the spots that were exposed to toxaphene, Wood commented that " . . . the control and treated fish may be identified on the basis of gill changes alone."[249] The lamellae were slender and delicate in the controls and clubbed and distally thickened in the exposed fish.

Juvenile coho and chinook salmon, fed a DDT-containing diet, showed localized degeneration of the distal convoluted tubule in the kidney. After prolonged feeding of the contaminated diet, the size of the liver decreased.[2] Electron microscopy revealed no drastic histopathological changes in the liver of guppies exposed to DDT up to 28 days. Zebra fish on the other hand, after 72 hr in 0.3 $\mu g/\ell$ or after 24 hr in 1 $\mu g/\ell$, lost glycogen completely, showed decreased cell size, and a cytoplasm filled with rough endoplasmic reticulum (Figure 1). The greater amount of lipid in the guppy seems responsible for its greater tolerance of DDT.[250] Acute pathological changes in the liver of *Channa punctata* were noticed 8 hr after an i.p. injection of endrin.[251] Subsequent to an acute exposure (96 hr) to endrin, swollen hepatic cells, liver cord disarray, necrosis, and vacuolation of cytoplasm were noticed in *Channa*.[252] [253] Sheepshead minnows exposed for 116 days to 0.78 $\mu g/\ell$ chlordecone had fatty, degenerated livers, with large vacuoles in hepatocytes.[20] Mirex-exposed cutthroat trout and goldfish showed pathological changes in the gills, with the adjacent lamellae being fused; edema in gills and hemorrhaged spaces were also observed. No such changes were observed in similarly treated bluegills.[254] Sheepshead minnows exposed to heptachlor (3.5 to 4.3 $\mu g/\ell$) showed various histopathological disorders like fatty livers and pancreas, intestines with lytic, frothy epithelium, and gills with edematous separation of lamellae,[255] not evident in the control fish. Edema in gill lamellae and sloughing off of the epithelium of the small intestine were reported in spot exposed to 1.35 μg dieldrin per liter for 4 days.[256]

Aroclor® 1216-exposed pinfish had severely vacuolated pancreatic exocrine tissue and basophilic and enlarged hepatocytes.[43] Fish given an i.p. injection of 50 mg of Aroclor® 1254 per kilogram body weight, had enlarged hepatocytes, with the endoplasmic reticulum appearing rounded rather than in parallel arrays.[257] Aroclor® 1254-exposed spot developed fatty livers.[258] In cod, fed a diet of herrings contaminated with Aroclor® 1254, severe histopathological disturbances like disorganization of the lobules, inhibition of spermatogenesis, and fatty necrosis were seen in the testes.[259] Not all changes caused by Aroclors® seem detrimental. The presence of 100 mg of Aroclor® 1254 per kilogram diet caused a significant reduction in the incidence of hepatic tumors in rainbow trout, caused by Aflatoxin B_1 (AFB$_1$).[260]

Sublethal concentrations of permethrin and also oral doses[261] caused histopathological changes in the gills but not in the liver and kidney of rainbow trout. An i.p. injec-

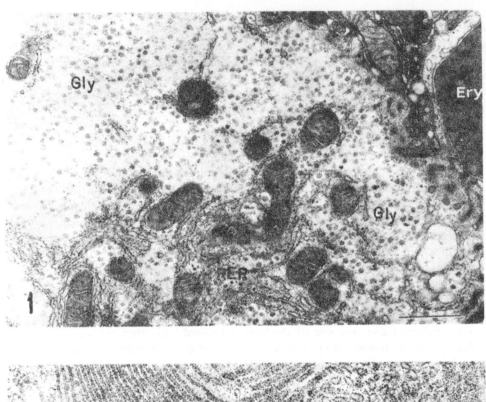

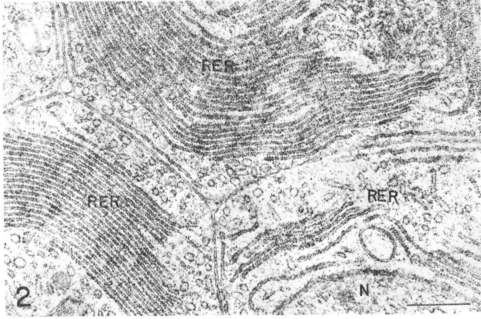

FIGURE 1.   Electron micrograph of DDT-induced changes in the hepatocytes of zebrafish. (1) Control fish showing glycogen granules (Gly) which fill most of the cytoplasmic compartment, along with rough endoplasmic reticulum (RER); an erythrocyte (Ery) is seen within a sinusoid. (2) Hepatocyte of zebrafish exposed to 1 μg/ℓ DDT for 24 hr. The cells are reduced in size and RER occupies most of the cytoplasmic compartment. N, nucleus × 16,745 (bar = 1 μm). (From Weis, P., *Chem. Biol. Interact.*, 8, 25, 1974. With permission.)

tion of 15 mg of methyl mercury per kilogram weight resulted in necrosis of renal tubule cells as well as exocrine pancreatic tissue.[262,263]

Sublethal concentrations of diazinon, methyl parathion, and dimethoate produced histopathological changes in the liver.[264,252,156] Bluegills collected 24 hr after treatment of ponds with 1 to 10 mg/l of 2,4-D showed pathological lesions.[230] Exposure of sockeye salmon fry to the butoxyethyl ester of 2,4-D for 96 hr (1 mg/l) resulted in hypertrophy and hyperplasia of the epithelial cells of the gills.[234]

Changes similar to pesticide-induced histopathological effects were observed in the livers of bluegills starved for 24 hr. Similarly, while there were distinct pathological changes in the livers of fish exposed to 0.1 mg PCP/l for 24 hr, those that were exposed for 96 hr had normal livers.[266] Hence, it is difficult to appreciate the real significance of pesticide-induced changes. Also, Sprague[265] underlined the lack of basic information on fish histology, which makes the interpretation of any observed change difficult.

VII. EFFECT OF PESTICIDES ON RESPIRATION

Respiratory distress is one of the early symptoms of pesticide poisoning. Exposure to sublethal concentrations is reported to increase respiratory activity, resulting in increased ventilation and hence, increased uptake of the toxicant. OCs have been reported to stimulate oxygen consumption at sublethal concentrations, and inhibit the oxygen uptake at lethal concentrations. In the case of OPs, a steady and progressive decline in oxygen consumption is noted. However, there have been exceptions and it is difficult to generalize.

Therapon exposed to DDT, dimethoate, and carbaryl consumed more oxygen than controls.[267] Exposure of *Saccobranchus* to lethal as well as sublethal concentrations resulted in increased opercular movement.[268] The oxygen consumption of white sucker exposed to 0.1 mg/l methoxychlor was 2 to 3.5 times that of the controls.[269] The oxygen consumption of suckers that died in 0.04 mg/l methoxychlor solution was similar to those exposed to 0.1 mg/l, whereas those that survived exposure to 0.04 mg/l had an oxygen consumption comparable to that of the controls. Rodgers and Beamish[270] observed a direct relationship between the uptake of methyl mercury by rainbow trout, and its oxygen consumption. Increasing concentrations of endosulfan decreased the oxygen consumption in *Macrognathus aculeatum*,[271] whereas oxygen consumption was first stimulated and then inhibited in *Labeo rohita*.[272] The oxygen consumption of *Labeo rohita* exposed to fenitrothion progressively decreased with increasing concentrations of the insecticide.[273] Acephate and fenitrothion at a concentration slightly higher than the 48-h LC 50 increased the ventilation rate and buccal amplitude.[61] A fourfold increase in the respiratory activity of herring embryos exposed to 2,4-DNP was noticed at the beginning of gastrulation. It fell to 50% of the control values at the time of closing of the blastopore, and decreased further until hatching.[177] Male guppies exposed to 0.5 or 1 mg/l sodium pentachlorophenate consumed more oxygen than the controls.[274]

Elevation of basal metabolic rate of rainbow trout exposed to permethrin was reported.[275] The pesticide-induced change in oxygen consumption of not only whole animals, but also of tissues was studied. Dichlorvos reduced the oxygen consumption of the excised tissues (gills, brain, muscle) of *Tilapia*.[276] In a study on the effect of herbicides with substituted acetic, propionic, and methyl groups on mitochondrial respiration, the number and position of the chloro-groups substituted on the phenoxy moiety influenced the oxygen uptake. This effect may be further increased or decreased by different phenoxy compounds containing different acid analogs.[277]

Several devices have been described to monitor the respiratory movements and rates in fish, which also have found use in assessing the impact of xenobiotic chemicals on fish.[278-280] A disposable electrode chamber for measuring opercular movements was reported.[281] A respirometer that stimulates the environment of a flowing water body was described.[282] This device showed that the ability of golden shiners to tolerate low oxygen tension was reduced by 50% following exposure to 5 mg/ℓ dibrom.[282]

VIII. EFFECT ON FEEDING AND GROWTH

A. Effect on Feeding

Reduced feeding and decreased ability to perceive the presence of food were evident in yellow bullheads exposed to pollutants (detergents).[283] After 96-hr exposure to metals, zebra danios required more time than the controls to consume a certain quantity of food.[284] When DDT was incorporated in the diet, young salmon found the food unpalatable, and consequently the food intake was markedly reduced.[285] After a 7-day exposure to tetrachloro dibenzo dioxin (TCDD), guppies showed a declining interest in food.[3] In the absence of contaminants, food odor was attractive to mullets, but when the food was contaminated with subacute levels of parathion, the fish avoided it.[286] *Mystus vittatus* exposed to sublethal concentrations of carbaryl for 27 days had significantly lower feeding and growth rates.[287] Exposure of coho salmon to Aroclor® 1254 or No. 2 fuel oil significantly reduced their prey-capturing ability.[288] On the other hand, exposure of the prey organism (grass shrimp) to mirex rendered them more susceptible to capture by pinfish.[289] While it is often reported that pesticides in general adversely affect feeding, Lingaraja et al.[1] noted that a predator fish, *Therapon jarbua*, upon being fed with sublethal doses of DDT, fed more frequently. *Therapon* either avoided DDT-exposed prey (mullets) or killed it, but did not consume it. When the predator was treated with DDT, it failed to discriminate between the treated and untreated prey.

B. Effect on Growth

The average weight of surviving cutthroat trout exposed orally to DDT, was higher than that of the controls. This was not the result of any stimulatory effect of DDT, but was due to the selective death of smaller and weaker fish.[232] Brook trout fed with sublethal doses of DDT were significantly longer than the controls,[290] and gained more weight than the controls.[291] In the latter study, untreated fish had a cumulative mortality of 1.2%, whereas fish exposed to 2 or 3 mg DDT per kilogram body weight had a cumulative mortality of 88.6 and 96.2%, respectively.[291] Bluegills collected from ponds treated with butoxyethyl ester of 2,4-D attained greater weights and lengths than the controls.[230] Growth of bluegills and shiners accelerated following the elimination of filamentous algae, and consequent redistribution of the nutrients following the treatment of ponds with diuron.[292] Increased growth, but no difference in the growth among five groups of minnows exposed to different doses of PCBs, was noted. The observed increased growth of this species was attributed to a disturbance of the hormonal system.[293] Growth of fathead minnows was stimulated by mirex at 13 μg/ℓ, whereas a concentration of 34 μg/ℓ had no effect.[23] A size-specific mortality in lake trout fry exposed to PCBs was reported; the affected fry were smaller.[294] Rainbow trout exposed to 35 ng PCP per liter had a higher growth than the controls, whereas those exposed to 660 ng/ℓ had a growth rate similar to that of the controls.[295] Similarly, exposure of goldfish[48] or rainbow trout[179] to endrin induced weight gain.

Exposure to pesticides, apart from resulting in stimulated growth or selective elimination of fish, also retarded or inhibited growth. Bluegills fed with mirex-contaminated diet (5 mg mirex per kilogram body weight at a rate of 5% of their body weight, 5 days a week, for 23 weeks) had retarded growth.[254]

Eels exposed to PCP had significantly less weight than the controls.[231] Growth and food conversion efficiency of sockeye salmon exposed to sublethal concentrations of sodium pentachlorophenate were greatly reduced. The 96-h LC 50 was 63 $\mu g/l$, while the EC 50 was 1.74 $\mu g/l$ for growth and 1.8 $\mu g/l$ for food conversion.[296] Toxaphene even at very low concentrations (39 or 68 ng/l) affected the growth of brook trout.[197,297] Growth, as determined by weight, was significantly reduced in brook trout fry exposed to sublethal concentrations of Aroclor® 1254 for 48 days. The increased use of vitamin C by the fish for detoxification of the xenobiotic chemical caused a functional deficiency of this vitamin. Since fish cannot synthesize vitamin C, this resulted in retardation of bone development.[298] Growth of lake trout fry was reduced at low concentrations of picloram or dinoseb.[299] Picloram,[300] malathion,[301] and Aroclor® 1262[302] reduced the growth of cutthroat trout, flagfish, and channel catfish, respectively. Fenitrothion, when administered orally, had no effect on the growth of rainbow trout.[303] The growth of fathead minnows exposed to PCP was retarded, which perhaps was the effect of the impurities (HCB, PCDDs, and PCDFs) present in commercial PCP.[5] Average standard length of fry, continuously exposed to heptachlor from fertilization, was reduced.[255]

IX. EFFECT OF PESTICIDES ON REPRODUCTION AND EARLY DEVELOPMENTAL STAGES OF FISH

It has been well said that in order to eliminate a species, a pollutant need not be lethal. The survival of a species is affected if its reproductive capacity is affected.[209] Usually the adult fish are less susceptible to the effect of xenobiotic organisms than the juveniles and larvae, which form the weakest link in the chain. Any alteration of the physiological or biochemical functioning of the younger developmental stages or reduction of their survival would affect the survival of the species itself.

A. Residues in Gonads and Gametes

The total DDT content of eggs of rainbow trout and *Acipenser* collected in Iran in 1974 was 256 and 91 to 1983 $\mu g/kg$, respectively.[304] Goldfish exposed to ^{14}C dieldrin accumulated the highest residue concentration in the testes.[305] In an analysis of fish from Smoky Hill River in Western Kansas, the gonads, especially the carp testes, tended to have the most residues. The residue levels in the testes were higher than in ovaries.[306] Bulkley et al.[307] considered that the small amount of milt shed by the male would not help it in reducing the body contaminant burden as much as spawning helps the female. On the other hand, Guiney et al.[308] reported that during spawning, the elimination of 2,5,2′,5′-tetrachlorobiphenyl through eggs and sperm was very rapid. A redistribution of the PCB residues of the different parts occurred during the process of germ cell maturation. After an i.p. injection of benzo(a)pyrene, a petroleum hydrocarbon, the concentration of the parent compound and its polar metabolites was 3 to 11 times higher in the ova and semen than in the ovaries and testes of English sole.[309] Testicular abnormalities like disorganization of lobules, inhibition of spermatogenesis, and necrosis were observed in cod that were fed a diet of herring containing Aroclor® 1254. The abnormalities were observed only in testes that contained rapidly proliferating cells or were functionally mature, but not in immature or regressed testes.[259]

B. Residues in Early Developmental Stages

In a pre-PCB era publication, DDT concentration of 668 $\mu g/kg$ was reported from the eggs of king salmon.[310] The total DDT residues in hatchery-reared trout eggs ranged from 64 to 1422 μg/kg and the loss during the 2 months following the swim-up stage amounted to 15 to 90%.[311] Trout eggs were reported to contain up to 1 ng dieldrin per gram weight.[312] When female *Fundulus* were exposed twice to 0.1 μg/l of ^{14}C-DDT for 24 hr, with a period of 24 hr in clean water in between the two treatments, the amount of DDT in eggs was 0.05 to 0.37 ng/g. As a consequence, although fertilization was not affected, the rate of development was markedly reduced; the DDT-containing eggs remained one or two stages behind the controls. In the last stages of embryonic development, the DDT-treated eggs caught up with the controls and hatching was normal.[313] Even 5 weeks after fertilization, juvenile sheepshead minnows retained 46% of the chlordane contained in the eggs at the time of fertilization.[314] It was estimated in 1967 that pinfish and Atlantic croakers in Pensacola Bay, Fla., carried about half a pound of DDT from the estuary and lost it in the process of spawning.[315]

C. Effects on Reproduction and Fecundity

At concentrations above the threshold toxicity, many organic compounds caused large pregnant females of *Gambusia affinis* to abort.[316] In experimental ponds, 2,4-D butylester at the highest concentrations tested (5 and 10 mg/l), caused a delay in the spawning of bluegills.[230] For those fish that attach their eggs to the aquatic vegetation, elimination of the weeds by the application of herbicides adversely affects their reproductive success.[317] Exposure of goldfish to endrin via food (143 or 430 μg/kg) affected the gametogenesis, as was evident by the lower gonadosomatic index (GSI) and smaller size of the testes than in controls. Exposure of steelhead trout sperm for > 30 min to 0.1 mg/l of methyl mercury affected their viability.[318] Following the experimental exposure of winter flounder to 1 μg/l DDT + 1 or 2 μg/l dieldrin, or 2 μg/l DDT until the residue load in the gonads reached those levels recorded in wild fish, the treated and control fish were artificially spawned. While 97.8% of the control eggs were fertilized, only 40% of the eggs containing 4.6 ng DDT per gram and 1.21 ng dieldrin per gram, and 0% of those eggs having 1.74 ng dieldrin per gram but no DDT, were fertilized.[319] Eggs of DDT-exposed adults showed abnormal gastrulation, and 39% of them had vertebral deformities upon hatching. The percent of deformed larvae was dose-dependent.[319] The insect chemosterilant tris(1-aziridinyl)phosphine oxide (TEPA) affected the male fertility at a concentration that was harmless to the females.[320] Exposure of carp to fenitrothion (0.3 and 1.5 mg/l) reduced steroidogenesis in the gonads.[321] Oviposition of *Oryzias latipes* was suppressed by 1 μg/l chlordecone after 292-hr exposure, or by 2 μg/l after 96-hr exposure.[322]

Following the feeding of DDT to brook trout at 0.5, 1, and 2 μg/kg/week, for 22 weeks, the fish that were fed lower doses had significantly higher number of ova, than those that were fed the highest dose.[290] Female fathead minnows exposed to 0.68 mg carbaryl per liter for 7 months, produced significantly less eggs.[323] Fenitrothion, at sublethal concentrations, affected egg production, whereas temephos at sublethal concentrations affected normal birth.[324] Fecundity and fertility of female sheepshead minnows exposed to 0.78 μg/l chlordecone was reduced.[20] Spawning and fecundity of fathead minnows exposed to 3, 13, or 34 μg/l mirex was reduced.

D. Effect on Hatching

Holden[325] reported that brown trout eggs containing aldrin (5 μg/kg) failed to hatch. Hatching of bluegill eggs exposed to a formulation of fenoprop (Kuron®) at 0.1, 1, 5, or 10 μg/l, was not affected, but all the fry hatching from eggs exposed to the highest concentration died.[326] On the contrary, hatchability of coho salmon eggs, exposed to PCBs, was much reduced. Exposure to PCB effectively reduced the embryonic development time; exposed eggs hatched 2 to 5 days earlier than the controls.[327] A single treatment of embryos of the Atlantic silverside with *p,p*'-DDT, malathion, or carbaryl induced optical malformations, although not dose-related. In treated groups the axis formation and heartbeat initiation were impaired. Microphthalmia (reduced size of eyes), unilateral or bilaterial anophthalmia (absence of eyes), and cyclopia (median eye) were some of the observed deformities. After hatching, lordotic fry were seen in the 10 μg/ℓ carbaryl or malathion, or 25 μg/ℓ DDT treatment groups. The three insecticides reduced the survival time of the embryos. These effects were observed even at concentrations that may occur temporarily in the environment.[328] Carbaryl, malathion, and parathion caused abnormalities in the circulatory system of the embryos of medaka.[329] Embryos, sacfry, and larvae of *Caranx* exposed to lindane had advanced hatching and larval characters.[330] Yolk utilization and sac absorption were also impaired. In the sacfry, scoliosis of the postanal region of the body, fin deformity, and variation in the length of the auditory vesicles were noticed. Hatching success of fish eggs exposed to low concentration of mirex increased.[23]

To determine whether the high DDT and PCB concentrations recorded in the eggs, fry, and adults of Lake Michigan in any way contributed to the reproductive failure of the species, the mortality of eggs (collected from Lake Michigan) was compared with that of hatchery-raised lake trout eggs and fry. Mortality of the former was high, i.e., 80 $\pm$ 16% in 1973 to 1974, compared with 46 $\pm$ 14% of the latter. No such difference was found either in 1974 to 1975, or in carefully conducted experiments. The authors concluded that the significant difference observed in 1973 to 1974 was attributable to causes other than the xenobiotic residues.[331] Ripe Baltic flounder, caught alive, were stripped of eggs or milt, artificially fertilized, and hatching success was studied. When the ovary residue concentration of PCB exceeded 120 ng/g, hatchability was affected. Other chlorinated hydrocarbons did not show any correlation between residue concentration and hatchability.[332] Hatching of fathead minnow eggs exposed to 0.31 μg chlordecone per liter was significantly lower than that of the controls.[23]

E. Survival of the Fry

In one of the first accounts on the effect of DDT on early developmental stages, Burdick et al.[333] found a definite correlation between the DDT concentration and the death of fry. The effect was first noticed in 1955 when 347,000 fry that hatched from Lake George lake trout eggs failed to survive. Fry from eggs collected in 1956 and 1957 also failed to survive. In 1958, the survival was negligible (0.9 to 1.4%) in two hatcheries that received the eggs. The fry died soon after hatching, had distended air bladders and intestines, and floated upside down near the surface, eventually sinking and dying. All the lots of eggs with a concentration higher than 2.95 mg DDT per kilogram died. Exposure of fertilized fish eggs to herbicides had no effect on the development of the fry, but affected the survival.[334] The mortality of the sacfry was higher when either or both of the gametes were obtained from the treated fish. Sacfry mortality was highest at the time of the absorption of the yolk, and the observed mortality seems the result of release of insecticides from fat.[290] Eggs and embryos of sheepshead minnows were not affected by exposure to 1 mg/ℓ endrin, but the fry died at much lower concentrations.[335] When the postembryonic development of rainbow trout eggs containing 2.7 mg/kg PCB and 0.09 mg/kg Σ DDT was followed, 60 to 70% of the fry were deformed 30 days after hatching. The deformities included scoliosis, lordosis, kyphosis, absence of caudal vertebrae, cranial deformities, and a projecting mandible. The deformed fry failed to absorb the yolk completely.[336]

After the control of parasitic lamprey which caused the disappearance of lake trout from Lake Michigan by mid-1950s, attempts were made to restock the lake with hatchery-reared lake trout fry. Despite the abundance of lake trout in the early 1970s, no naturally produced lake trout yearlings were ever found. It was suspected that the high levels of DDT and PCB residues in water and plankton of Lake Michigan might have contributed to the reproductive failure of lake trout. Fry hatched from eggs of lake trout from Lake Michigan were exposed for 6 months to PCB and DDE at concentrations similar to those found in offshore waters (1 ng/l DDE and 10 ng/l PCB in water, and 1 μg/g PCB and 0.1 μg/g DDE in food). Cumulative mortality of such exposed fry was nearly twice that of the control fry, although there was no significant difference in the growth, swimming performance, predator avoidance, temperature avoidance, and metabolism of the fry. Hence, the levels of PCB and DDE obtained in the water and biota of Lake Michigan during early 1970 would have been sufficient to cause significant reduction in the survival of Lake Mighigan trout fry that might have hatched in the lake.

Mortality of the young, born 6 months after the exposure of male and female *Poecilia* to 2,4,6-trichlorophenol, was higher than that of the control fry in spite of the former being maintained in clean water.[337] Susceptibility of younger life stages, resulting from exposed eggs or parents, was reported.[337,338]

X. EFFECT ON FISH-FOOD ORGANISMS AND COMMUNITY STRUCTURE

A. Effect on Fish-Food Organisms

1. Fish and Fish-Food Organisms

In assessing the toxicity of pesticides to fish, not only directly, but indirectly, the effect of pesticides on fish-food organisms needs to be evaluated.[339] In streams that received runoff from apple orchards, the diversity of taxa and the abundance of aquatic macroinvertebrates were severely affected.[340] Fish fed on oysters which had earlier accumulated high concentrations of DDT from ambient water died within 2 days.[341] In an ecological survey of the biota of Lake Poinsett, in South Dakota, highest concentrations of Σ DDT were found in aquatic invertebrates.[342] Many invertebrates were reportedly more sensitive than fish to pesticides.[343]

Following the aerial spraying of forests in the Boulder River drainage in Montana, large-scale mortality and drift of aquatic insects in the adjoining streams were noticed.[344] Similar drift of aquatic insects was reported after the spraying of methoxychlor[345,346] and carbaryl.[347] Following the spraying of fenitrothion, salmon population did not recover for 6 years, possibly because of the scarcity of food, among other factors.[348] Salmon fry and parr were feeding mostly on dipteran, ephemeropteran, and trichopteran larvae before the spraying of New Brunswick forests with DDT. Because of the disappearance of these larvae after spraying, the fry had to subsist on chironomid larvae and the parr on dipterans, snails, and worms to makeup for the loss of normal diet. It was only after 3 to 5 years that the representative insect larval fauna that constituted the usual diet of salmon fry and parr returned to normal levels.[349,350] Following an aerial application of acephate on the forests of Maine, brook trout were found feeding for 2 days on terrestrial insects, beetles, moths, spiders, and wasps.[110] Shrimp and mayfly larvae disappeared following the aerial spraying of a Nigerian savanna with pyrethroids.[351]

Contamination of Chew Valley Lake and its feeder stream in England with residues arising from sheepdips, killed or prevented the reproduction of daphnids and other fish-food organisms.[352] Experimental spraying of farm ponds with Dursban® [104,353] and parathion[354,355] at recommended rates for mosquito control, resulted in the drastic reduction of the aquatic insect populations that formed the major source of food for bluegills. A marked reduction of cladocerans and copepod nauplii followed the application of diflubenzuron to five experimental ponds. For 1 month after application, black guppies and brown bullheads had to feed on an altered diet.[356]

2. Toxicity of Pesticides to Aquatic Invertebrates

Many of the aquatic invertebrates, especially cladocerans, copepods, and insect larvae that form an important source of food for many species of fish, are reported to be more sensitive to pesticides than fish.[357,358] Many herbicides, at the rates applied in the field, were reported to result in concentrations much higher than the 48-h EC 50 to *Daphnia*.[359,360] Very low concentrations of pesticides are reported to be extremely toxic to the larvae and early juveniles of many crustaceans, especially at the time of molting.[361] Mayflies, which constitute an important link in the food chain and represent one of the quickest and most efficient groups that convert detritus into high quality fish-food,[362] are extremely sensitive to even low concentrations of pesticides.[363] The pyrethroids, especially the cyano-substituted ones, viz., cypermethrin and fenvalerate, were more toxic to aquatic invertebrates than to fish.[364] Increased incidence of cannibalism among stoneflies in experimental tanks (exposed to DDT) was observed. Further, pesticide-exposed individuals were hyperactive and crawled out into the open, whereas control stoneflies always remained hidden under rocks.[365] Such altered behavior in nature will surely affect the survival of the populations.

B. Effect of Pesticides on Community Structure

The long-range effects of pesticides on community structure and function are subtle and little understood. Hitherto, toxicity testing has been confined to single species or rarely to an assemblage of a few species in a test tank or an experimental pond. It is only recently that it has been realized that pollutants may exert an overall modifying effect on the communities.[366-370] Heckman[371] recently described in detail the changes in the aquatic communities associated with the orchard ditches near Hamburg that took place as a result of 25 years of intense pesticide use. In one of the rare studies of its type, he compared the community structure obtained in 1978 with that of 1953. Certain species, especially predators, had completely disappeared, which stimulated the growth of other species. Several species became resistant to agricultural chemicals.

Even if fish are not directly affected by pesticides, any change of the community structure is bound to affect them. As pointed out by Swingle,[372] aquatic environments may appear to have recovered from the effect of pollutants, especially if judged by the physicochemical conditions, and such environments may support a good standing crop of fish, but the actual population analysis often shows a reduction in desirable and harvestable species and increase in the number of species that cannot be exploited.

XI. CONCLUSIONS

Pesticides have been reported to cause many alterations in fish, ranging from change of color and alteration of behavior to damage to the gonads and reduction in the survival of offspring. Many of the reported changes, however, are not dose-dependent, and often are observed only in laboratory experiments and not in the field. Further, in the case of biochemical changes, results of in vitro studies often differed from those of in vivo studies. Hence, the ecological significance of all the reported pesticide-induced changes is hard to evaluate. Behavioral tests, biochemical tests, histopathological, and other types of studies may serve a useful purpose when combined with other tests, but as has been pointed out by Brungs and Mount[373] and Mehrle and Mayer,[198] many of these tests are rated very low in any scheme of environmental hazard evaluation, because they are subjective. The extent to which the observed change may impair the functioning of an organism or a species is difficult to judge. In terms of present utility, acute toxicity tests followed by embryo-larval tests and chronic life cycle tests are ranked highest; histological tests are ranked 9th, and physiological and biochemical tests 10th out of 11 tests.[198]

REFERENCES

1. Lingaraja, T., Rao, P. S. S., and Venugopalan, V. K., DDT induced ethological changes in estuarine fish, *Environ. Biol. Fish.*, 4, 83, 1979.
2. Buhler, D. R., Rasmusson, M. E., and Shanks, W. E., Chronic oral DDT toxicity in juvenile coho and chinook salmon, *Toxicol. Appl. Pharmacol.*, 14, 535, 1969.
3. Norris, L. A. and Miller, R. A., The toxicity of 2,3,7,8-tetrachlorodibenzo-p-dioxin (TCDD) in guppies (*Poecilia reticulatus* Peters), *Bull. Environ. Contam. Toxicol.*, 12, 76, 1974.
4. Holcombe, G. W., Phipps, G. L., and Tanner, D. K., The acute toxicity of kelthane, Dursban, disulfoton, pydrin and permethrin to fathead minnows *Pimephales promelas* and rainbow trout *Salmo gairdneri*, *Environ. Pollut.*, 29, 167, 1982.
5. Cleveland, L., Buckler, D. R., Mayer, F. L., and Branson, D. R., Toxicity of three preparations of pentachlorophenol to fathead minnows — a comparative study, *Environ. Toxicol. Chem.*, 1, 205, 1982.
6. Symons, P. E. K., Behaviour of young Atlantic salmon (*Salmo salar*) exposed to or force-fed fenitrothion, an organophosphate insecticide, *J. Fish. Res. Board Can.*, 30, 651, 1973.
7. Leung, T. S., Naqvi, S. M., and Leblanc, C., Toxicities of two herbicides (basagran, diquat) and an algicide (cutrine-plus) to mosquitofish *Gambusia affinis*, *Environ. Pollut.*, 30, 153, 1983.
8. Palawski, D., Buckler, D. R., and Mayer, F. L., Survival and condition of rainbow trout (*Salmo gairdneri*) after acute exposures to methyl parathion, triphenyl phosphate, and DEF, *Bull. Environ. Contam. Toxicol.*, 30, 614, 1983.
9. Henderson, C., Pickering, Q. H., and Tarzwell, C. M., Relative toxicity of ten chlorinated hydrocarbon insecticides to four species of fish, *Trans. Am. Fish. Soc.*, 88, 23, 1959.
10. Klaverkamp, J. F., Duangsawasdi, M., Macdonald, W. A., and Majewski, H. S., An evaluation of fenitrothion toxicity in four life stages of rainbow trout, *Salmo gairdneri, in Aquatic Toxicology and Hazard Evaluation*, ASTM STP 634, Mayer, F. L. and Hamelink, J. L., Eds., American Society for Testing and Materials, Philadelphia, 1977, 231.
11. Chin, Y. N. and Sudderuddin, K. I., Effect of methamidophos on the growth rate and esterase activity of the common carp *Cyprinus carpio* L., *Environ. Pollut.*, 18, 213, 1979.
12. Mayer, F. L., Adams, W. J., Finley, M. T., Michael, P. R., Mehrle, P. M., and Seger, V. M., Phosphate ester hydraulic fluids: an aquatic environmental assessment of pydrauls 50E and 115E, in *Aquatic Toxicology and Hazard Assessment: fourth Conference*, ASTM STP 737, Branson, D. R. and Dickson, K. L., Eds., American Society for Testing and Materials, Philadelphia, 1981, 103.
13. Rosenthal, H. and Stelzer, R., Wirkungen von 2,4- and 2,5-dinitrophenol and die Embryonalentwicklung des Herings *Clupea harengus, Mar. Biol.*, 5, 325, 1970.
14. Weis, J. S. and Weis, P., Retardation of fin regeneration in *Fundulus* by several insecticides, *Trans. Am. Fish. Soc.*, 104, 135, 1975.
15. Weis, P. and Weis, J. S., Congenital abnormalities in estuarine fishes produced by environmental contaminants, in *Animals as Monitors of Environmental Pollutants*, National Academy of Sciences, Washington, D. C., 1979, 94.
16. Darsie, R. F., Jr. and Corriden, F. E., The toxicity of malathion to killifish (Cyprinodontidae) in Delaware, *J. Econ. Entomol.*, 52, 696, 1959.
17. McCann, J. A. and Jasper, R. L., Vertebral damage to bluegills exposed to acutely toxic levels of pesticides, *Trans. Am. Fish. Soc.*, 101, 317, 1972.
18. Weis, P. and Weis, J. S., Abnormal locomotion associated with skeletal malformations in the sheepshead minnow, *Cyprinodon variegatus*, exposed to malathion, *Environ. Res.*, 12, 196, 1976.
19. Couch, J. A., Winstead, J. T., and Goodman, L. R., Kepone-induced scoliosis and its histological consequences in fish, *Science*, 197, 585, 1977.
20. Goodman, L. R., Hansen, D. J., Manning, C. S., and Faas, L. F., Effects of Kepone® on the sheepshead minnow in an entire life-cycle toxicity test, *Arch. Environ. Contam. Toxicol.*, 11, 335, 1982.
21. Couch, J. A., Winstead, J. T., Hansen, D. J., and Goodman, L. R., Vertebral dysplasia in young fish exposed to the herbicide trifluralin, *J. Fish. Dis.*, 2, 35, 1979.
22. Wells, D. E. and Cowan, A. A., Vertebral dysplasia in salmonids caused by the herbicide trifluralin, *Environ. Pollut.*, 29, 249, 1982.
23. Buckler, D. R., Witt, A., Jr., Mayer, F. L., and Huckins, J. N., Acute and chronic effects of Kepone and mirex on the fathead minnow, *Trans. Am. Fish. Soc.*, 110, 270, 1981.
24. Mehrle, P. M., Haines, T. A., Hamilton, S., Ludke, L., Mayer, F. L., and Ribick, M. A., Relationship between body contaminants and bone development in east-coast striped bass, *Trans. Am. Fish. Soc.*, 111, 231, 1982.

25. Bengtsson, B.-E., Vertebral damage in fish induced by pollutants, in *Sublethal Effects of Toxic Chemicals on Aquatic Animals,* Koeman, J. H. and Strik, J. J. T. W. A., Eds., Elsevier, Amsterdam, 1975, 23.

26. Warner, R. E., Peterson, K. K., and Borgman, L., Behavioural pathology in fish: a quantitative study of sublethal pesticide toxication, *J. Appl. Ecol.,* 3, 223, 1966.

27. Scherer, E., Avoidance of fenitrothion by goldfish *(Carassius auratus), Bull. Environ. Contam. Toxicol.,* 13, 492, 1975.

28. Bull, J. and McInerney, J. E., Behavior of juvenile coho salmon (*Oncorhynchus kisutch*) exposed to sumithion (fenitrothion), an organophosphate insecticide, *J. Fish. Res. Board Can.,* 31, 1867, 1974.

29. Nevins, M. J. and Johnson, W. W., Acute toxicity of phosphate ester mixtures to invertebrates and fish, *Bull. Environ. Contam. Toxicol.,* 19, 250, 1978.

30. Chliamovitch, Y.-P. and Kuhn, C., Behavioural, haematological and histological studies on acute toxicity of bis(tri-n-butyltin)oxide on *Salmo gairdneri* Richardson and *Tilapia rendalli* Boulenger, *J. Fish. Biol.,* 10, 575, 1977.

31. Hansen, D. J., Avoidance of pesticides by untrained sheepshead minnows, *Trans. Am. Fish. Soc.,* 98, 426, 1969.

32. Hansen, D. J., Matthews, E., Nall, S. L., and Dumas, D. P., Avoidance of pesticides by untrained mosquitofish, *Gambusia affinis, Bull. Environ. Contam. Toxicol.,* 8, 46, 1972.

33. Sprague, J. B. and Drury, D. E., Avoidance reactions of salmonid fish to representative pollutants, *Adv. Water Pollut. Res.,* 1, 169, 1969.

34. Matton, P. and LaHam, Q. N., Effect of the organophosphate Dylox on rainbow trout larvae, *J. Fish. Res. Board Can.,* 26, 2193, 1969.

35. Anderson, J. M. and Peterson, M. R., DDT: sublethal effects on brook trout nervous system, *Science,* 164, 440, 1969.

36. Rand, G. M., The effect of exposure to a subacute concentration of parathion on the general locomotor behavior of the goldfish, *Bull. Environ. Contam. Toxicol.,* 18, 259, 1977.

37. Dodson, J. J. and Mayfield, C. I., Modification of the rheotropic response of rainbow trout (*Salmo gairdneri)* by sublethal doses of the aquatic herbicides diquat and simazine, *Environ. Pollut.,* 18, 147, 1979.

38. Hildebrand, L. D., Sullivan, D. S., and Sullivan, Th. P., Experimental studies of rainbow trout populations exposed to field applications of Roundup® herbicide, *Arch. Environ. Contam. Toxicol.,* 11, 93, 1982.

39. Abban, E. K. and Samman, J., Preliminary observations on the effect of the insect larvicide Abate on fish catches in the river Oti, Ghana, *Environ. Pollut.,* 21, 307, 1980.

40. Folmar, L. C., Overt avoidance reaction of rainbow trout fry to nine herbicides, *Bull. Environ. Contam. Toxicol.,* 15, 509, 1976.

41. Saunders, J. W., Mass mortalities and behavior of brook trout and juvenile Atlantic salmon in a stream polluted by agricultural pesticides, *J. Fish. Res. Board Can.,* 26, 695, 1969.

42. Dill, P. A. and Saunders, R. C., Retarded behavioral development and impaired balance in Atlantic salmon (*Salmo salar)* alevins hatched from gastrulae exposed to DDT, *J. Fish. Res. Board Can.,* 31, 1936, 1974.

43. Hansen, D. J., Parrish, P. R., and Forester, J., Aroclor 1016: toxicity to and uptake by estuarine animals, *Environ. Res.,* 7, 363, 1971.

44. Adelman, I. R., Smith, L. L., Jr., and Siesennop, G. D., Acute toxicity of sodium chloride, pentachlorophenol, Guthion®, and hexavalent chromium to fathead minnows (*Pimephales promelas)* and goldfish *(Carassius auratus), J. Fish. Res. Board Can.,* 33, 203, 1976.

45. Lehtinen, K. J. and Oikari, A., Sublethal effects of kraft pulp mill waste water on the perch, *Perca fluviatilis,* studies by rotary-flow and histological techniques, *Ann. Zool. Fenn.,* 17, 255, 1980.

46. Abede, Z. H. and McKinley, W. P., Bioassay of captan by zebrafish larvae, *Nature,* 216, 1321, 1967.

47. Anderson, J. M., Effect of sublethal DDT on the lateral line of brook trout, *Salvelinus fontinalis, J. Fish. Res. Board Can.,* 25, 2677, 1968.

48. Grant, B. F. and Mehrle, P. M., Chronic endrin poisoning in goldfish, *Carassius auratus, J. Fish. Res. Board Can.,* 27, 2225, 1970.

49. Argyle, R. L., Williams, G. C., and Dupree, H. K., Endrin uptake and release by fingerling channel catfish *(Ictalurus punctatus), J. Fish. Res. Board Can.,* 30, 1742, 1973.

50. Anderson, P. D. and Weber, L. J., Toxic response as a quantitative function of body size, *Toxicol. Appl. Pharmacol.,* 33, 471, 1975.

51. Ellgaard, E. G., Ochsner, J. C., and Cox, J. K., Locomotor hyperactivity induced in the bluegill sunfish, *Lepomis macrochirus,* by sublethal concentrations of DDT, *Can. J. Zool.,* 55, 1077, 1977.

52. Thatcher, T. O., A morphological defect in shiner perch resulting from chronic exposure to chlorinated sea water, *Bull. Environ. Contam. Toxicol.,* 21, 433, 1979.

53. Weis, P. and Weis, J. S., Sevin and schooling behavior of *Menidia menidia* in the presence of the insecticide Sevin (carbaryl), *Mar. Biol.*, 28, 261, 1974.

54. Zuyev, G. V. and Belyayev, V. V., An experimental study of the swimming of fish in groups as exemplified by the horsemackerel (*Trachurus mediterraneus*, Ponticus, Aleev), *J. Ichthyol.*, 10, 545, 1970; as cited in Weis, P. and Weis, J. S., Sevin and schooling behavior of *Menidia menidia* in the presence of the insecticide Sevin (carbaryl), *Mar. Biol.*, 28, 261, 1974.

55. Saunders, R. L. and Sprague, J. B., Effects of copper-zinc mining pollution on a spawning migration of Atlantic salmon, *Water Res.*, 1, 419, 1967.

56. Elson, P. F., Lauzier, L. N., and Zitko, V., A preliminary study of salmon movements in a polluted estuary, in *Marine Pollution and Sea Life*, Ruivo, M., Ed., Unipub, New York, 1972.

57. Carlson, R. W., Some characteristics of ventilation and coughing in the bluegill *Lepomis macrochirus* Rafinesque, *Environ. Pollut.*, 29, 35, 1982.

58. Carlson, R. W. and Drummond, R. A., Fish cough response — a method for evaluating quality of treated complex effluents, *Water Res.*, 12, 1, 1978.

59. Schaumburg, F. D., Howard, T. E., and Walden, C. C., A method to evaluate the effects of water pollutants on fish respiration, *Water Res.*, 1, 731, 1967.

60. Drummond, R. A., Spoor, W. A., and Olson, G. F., Some short-term indicators of sublethal effects of copper on brook trout, *Salvelinus fontinalis*, *J. Fish. Res. Board Can.*, 30, 698, 1973.

61. Duangsawasdi, M. and Klaverkamp, J. F., Acephate and fenitrothion toxicity in rainbow trout: effects of temperature stress and investigations on the sites of action, in *Aquatic Toxicology*, ASTM STP 667, Marking, L. L. and Kimerle, R. A., Eds., American Society for Testing and Materials, Philadelphia, 1979, 35.

62. Aubin, A. E. and Johansen, P. H., The effects of an acute DDT exposure on the spontaneous electrical activity of goldfish cerebellum, *Can. J. Zool.*, 47, 163, 1969.

63. Anderson, J. M. and Prins, H. B., Effects of sublethal DDT on a simple reflex in brook trout, *J. Fish. Res. Board Can.*, 27, 331, 1970.

64. Davy, F. B., Kleerekoper, H., and Gensler, P., Effects of exposure to sublethal DDT on the locomotor behavior of the goldfish (*Carassius auratus*), *J. Fish. Res. Board Can.*, 29, 1333, 1972.

65. Hatfield, C. T. and Johansen, P. H., Effects of four insecticides on the ability of Atlantic salmon parr (*Salmo salar*) to learn and retain a simple conditioned response, *J. Fish. Res. Board Can.*, 29, 315, 1972.

66. McNicholl, P. G. and Mackay, W. C., Effect of DDT and M.S.222 on learning a simple conditioned response in rainbow trout (*Salmo gairdneri*), *J. Fish. Res. Board Can.*, 32, 661, 1975.

67. Klaverkamp, J. F., Lockhart, W. L., Metner, D., and Grift, N., Effects of chronic DDT/DDE exposure on anesthetic induction and recovery times in rainbow trout (*Salmo gairdneri*), *J. Fish. Res. Board Can.*, 33, 1331, 1976.

68. Hatfield, C. T. and Anderson, J. M., Effects of two insecticides on the vulnerability of Atlantic salmon (*Salmo salar*) parr to brook trout (*Salvelinus fontinalis*) predation, *J. Fish. Res. Board Can.*, 29, 27, 1972.

69. Farr, J. A., Impairment of antipredator behavior in *Palaemonetes pugio* by exposure to sublethal doses of parathion, *Trans. Am. Fish. Soc.*, 106, 287, 1977.

70. Schneider, M. J., Barraclough, S. A., Genoway, R. G., and Wolford, M. L., Effects of phenol on predation of juvenile rainbow trout *Salmo gairdneri*, *Environ. Pollut.*, 23, 121, 1980.

71. Colgan, P. W., Cross, J. A., and Johansen, P. H., Guppy behavior during exposure to a sub-lethal concentration of phenol, *Bull. Environ. Contam. Toxicol.*, 28, 20, 1982.

72. Ogilvie, D. M. and Anderson, J. M., Effect of DDT on temperature selection by young Atlantic salmon, *Salmo salar*, *J. Fish. Res. Board Can.*, 22, 503, 1965.

73. Miller, D. L. and Ogilvie, D. M., Temperature selection in brook trout (*Salvelinus fontinalis*) following exposure to DDT, PCB or phenol, *Bull. Environ. Contam. Toxicol.*, 14, 545, 1975.

74. Gardner, D. R., The effect of some DDT and methoxychlor analogs on temperature selection and lethality in brook trout fingerlings, *Pestic. Biochem. Physiol.*, 2, 437, 1973.

75. Peterson, R. H., Temperature selection of Atlantic salmon (*Salmo salar*) and brook trout (*Salvelinus fontinalis*) as influenced by various chlorinated hydrocarbons, *J. Fish. Res. Board Can.*, 30, 1091, 1973.

76. Domanik, A. M. and Zar, J. H., The effect of malathion on the temperature selection response of the common shiner, *Notropis cornutus* (Mitchill), *Arch. Environ. Contam. Toxicol.*, 7, 193, 1978.

77. Ogilvie, D. M. and Miller, D. L., Duration of a DDT-induced shift in the selected temperature of Atlantic salmon (*Salmo salar*), *Bull. Environ. Contam. Toxicol.*, 16, 96, 1976.

78. Peterson, R. H., Temperature selection of juvenile Atlantic salmon (*Salmo salar*), as influenced by various toxic substances, *J. Fish. Res. Board Can.*, 33, 1722, 1976.

79. Hansen, D. J., DDT and malathion: effect on salinity selection by mosquitofish, *Trans. Am. Fish. Soc.*, 2, 346, 1972.

80. Spoor, W. A., Neiheisel, T. W., and Drummond, R. A., An electrode chamber for recording respiratory and other movements of free-swimming animals, *Trans. Am. Fish. Soc.*, 100, 22, 1971.

81. Rommel, S. A., Jr., A simple method for recording fish heart and operculum beats without the use of implanted electrodes, *J. Fish. Res. Board Can.*, 30, 693, 1973.

82. Scherer, E. and Nowak, S., Apparatus for recording avoidance movements of fish, *J. Fish. Res. Board Can.*, 30, 1594, 1973.

83. Cripe, C. R., An automated device (ACARS) for studying avoidance of pollutant gradients by aquatic organisms, *J. Fish. Res. Board Can.*, 36, 11, 1979.

84. Cairns, J., Jr., Thompson, K. W., Landers, J. D., Jr., McKee, M. J., and Hendricks, A. C., Suitability of some freshwater and marine fishes for use with a minicomputer interfaced biological monitoring system, *Water Resour. Bull.*, 16, 421, 1980.

85. O'Brien, R. D., *Insecticides: Action and Metabolism*, Academic Press, New York, 1967, 1.

86. Weiss, C. M., The determination of cholinesterase in the brain tissue of three species of freshwater fish and its inactivation in vivo, *Ecology*, 39, 194, 1958.

87. Weiss, C. M., Response of fish to sub-lethal exposures of organic phosphorus insecticides, *Trans. Am. Fish. Soc.*, 31, 580, 1959.

88. Weiss, C. M., Physiological effect of organic phosphorus insecticides on several species of fish, *Trans. Am. Fish. Soc.*, 90, 143, 1961.

89. Weiss, C. M. and Gakstatter, J. H., Detection of pesticides in water by biochemical assay, *J. Water Pollut. Cont. Fed.*, 36, 240, 1964.

90. Williams, A. K. and Sova, C. R., Acetylcholinesterase levels in brains of fishes from polluted waters, *Bull. Environ. Contam. Toxicol.*, 1, 198, 1966.

91. Coppage, D. L. and Matthews, E., Short-term effects of organophosphate pesticides on cholinesterases of estuarine fishes and pink shrimp, *Bull. Environ. Contam. Toxicol.*, 11, 483, 1974.

92. Coppage, D. L., Organophosphate pesticides: specific level of brain AChE inhibition related to death in sheepshead minnows, *Trans. Am. Fish. Soc.*, 101, 534, 1972.

93. Coppage, D. L., Matthews, E., Cook, G. H., and Knight, J., Brain acetylcholinesterase inhibition in fish as a diagnosis of environmental poisoning by malathion, O,O-dimethyl S-(1,2-dicarbethoxyethyl) phosphorodithioate, *Pestic. Biochem. Physiol.*, 5, 536, 1975.

94. Gibson, J. R. and Ludke, J. L., Effect of SKF-525A on brain acetylcholinesterase inhibition by parathion in fishes, *Bull. Environ. Contam. Toxicol.*, 9, 140, 1973.

95. Coppage, D. L. and Matthews, E., Brain-acetylcholinesterase inhibition in marine teleost during lethal and sublethal exposures to 1,2-dibromo-2,2-dichloroethyl dimethyl phosphate (naled) in sea water, *Toxicol. Appl. Pharmacol.*, 31, 128, 1975.

96. Coppage, D. L. and Braidech, T. E., River pollution by anticholinesterase agents, *Water Res.*, 10, 19, 1976.

97. Cook, G. H., Moore, J. C., and Coppage, D. L., The relationship of malathion and its metabolites to fish poisoning, *Bull. Environ. Contam. Toxicol.*, 16, 283, 1976.

98. Holland, T., Coppage, D. L., and Butler, P. A., Use of fish brain acetylcholinesterase to monitor pollution by organophosphorus pesticides, *Bull. Environ. Contam. Toxicol.*, 2, 156, 1967.

99. Gibson, J. R., Ludke, J. L., and Ferguson, D. E., Sources of error in the use of fish-brain acetylcholinesterase activity as a monitor for pollution, *Bull. Environ. Contam. Toxicol.*, 4, 17, 1969.

100. Hogan, J. W., Water temperature as a source of variation in specific activity of brain acetylcholinesterase of bluegills, *Bull. Environ. Contam. Toxicol.*, 5, 347, 1970.

101. Coppage, D. L., Characterization of fish brain acetylcholinesterase with an automated pH stat for inhibition studies, *Bull. Environ. Contam. Toxicol.*, 6, 304, 1971.

102. Benke, G. M., Cheever, K. L., Mirer, F. E., and Murphy, S. D., Comparative toxicity, anticholinesterase action and metabolism of methyl parathion and parathion in sunfish and mice, *Toxicol. Appl. Pharmacol.*, 28, 97, 1974.

103. Benke, G. M. and Murphy, S. D., Anticholinesterase action of methyl parathion, parathion and azinphosmethyl in mice and fish: onset and recovery of inhibition, *Bull. Environ. Contam. Toxicol.*, 12, 117, 1974.

104. Macek, K. J., Walsh, D. F., Hogan, J. W., and Holz, D. D., Toxicity of the insecticide Dursban® to fish and aquatic invertebrates in ponds, *Trans. Am. Fish. Soc.*, 101, 420, 1972.

105. Thirugnanam, M. and Forgash, A. J., Environmental impact of mosquito pesticides: toxicity and anticholinesterase activity of chlorpyrifos to fish in a salt marsh habitat, *Arch. Environ. Contam. Toxicol.*, 5, 415, 1977.

106. Verma, S. R., Tyagi, A. K., Bhatnagar, M. C., and Dalela, R. C., Organophosphate poisoning to some freshwater teleosts-acetylcholinesterase inhibition, *Bull. Environ. Contam. Toxicol.*, 21, 502, 1979.

107. Kanazawa, J., *In vitro* and *In vivo* effects of organophosphorus and carbamate insecticides on brain acetylcholinesterase activity of freshwater fish, topmouth gudgeon, *Bull. Natl. Inst. Agric. Sci. Ser. C*, 37, 19, 1983.

108. Post, G. and Leasure, R. A., Sublethal effect of malathion to three salmonid species, *Bull. Environ. Contam. Toxicol.*, 12, 312, 1974.

109. Kabeer Ahammad Sahib, I. and Ramana Rao, K. V., Correlation between subacute toxicity of malathion and acetylcholinesterase inhibition in the tissues of the teleost *Tilapia mossambica*, *Bull. Environ. Contam. Toxicol.*, 24, 711, 1980.

110. Rabeni, C. F. and Stanley, J. G., Operational spraying of acephate to suppress spruce budworm has minor effects on stream fishes and invertebrates, *Bull. Environ. Contam. Toxicol.*, 23, 327, 1979.

111. Olson, D. L. and Christensen, G. M., Effects of water pollutants and other chemicals on fish acetylcholinesterase (in vitro), *Environ. Res.*, 21, 327, 1980.

112. Koch, R. B., Inhibition of animal tissue ATPase activities by chlorinated hydrocarbon pesticides, *Chem. Biol. Interact.*, 1, 199, 1970.

113. Koch, R. B., Chlorinated hydrocarbon insecticides: inhibition of rabbit brain ATPase activities, *J. Neurochem.*, 16, 269, 1969.

114. Cutkomp, L. K., Koch, R. B., and Desaiah, D., Inhibition of ATPases by chlorinated hydrocarbons, in *Insecticide Mode of Action*, Coats, J. R., Ed., Academic Press, New York, 1982, 45.

115. Matsumura, F. and Patil, K. C., Adenosine triphosphate sensitive to DDT in synapses of rat brain, *Science*, 166, 121, 1969.

116. Matsumura, F., Bratkowski, T. A., and Patil, K. C., DDT: inhibition of an ATPase in the rat brain, *Bull. Environ. Contam. Toxicol.*, 4, 262, 1969.

117. Ghiasuddin, S. M. and Matsumura, F., DDT inhibition of Ca-ATPase of the peripheral nerves of the American lobster, *Pestic. Biochem. Physiol.*, 10, 151, 1979.

118. Matsumura, F. and Ghiasuddin, S. M., Characteristics of DDT-sensitive Ca-ATPase in the axonic membrane, in *Neurotoxicology of Insecticides and Pheromones*, Narahashi, Ed., Plenum Press, New York, 1979, 245.

119. Koch, R. B., Inhibition of oligomycin sensitive and insensitive fish adenosine triphosphatase activity by chlorinated hydrocarbon insecticides, *Biochem. Pharmacol.*, 20, 3243, 1971.

120. Davis, P. W. and Wedemeyer, G. A., Na$^+$, K$^+$-activated- ATPase inhibition in rainbow trout: a site for organochlorine pesticide toxicity?, *Comp. Biochem. Physiol.*, 408, 823, 1971.

121. Koch, R. B., Desaiah, D., Yap, H. H., and Cutkomp, L. K., Polychlorinated biphenyls: effect of long-term exposure on ATPase activity in fish, *Pimephales promelas*, *Bull. Environ. Contam. Toxicol.*, 7, 87, 1972.

122. Davis, P. W., Friedhoff, J. M., and Wedemeyer, G. A., Organochlorine insecticide, herbicide and polychlorinated biphenyl (PCB) inhibition of NaK-ATPase in rainbow trout, *Bull. Environ. Contam. Toxicol.*, 8, 69, 1972.

123. Campbell, R. D., Leadem, T. P., and Johnson, D. W., The *in vivo* effect of *p,p'* DDT on Na$^+$-K$^+$-activated ATPase activity in rainbow trout *(Salmo gairdneri)*, *Bull. Environ. Contam. Toxicol.*, 11, 425, 1974.

124. Desaiah, D. and Koch, R. B., Toxaphene inhibition of ATPase activity in catfish, *Ictalurus punctatus*, tissues, *Bull. Environ. Contam. Toxicol.*, 13, 238, 1975.

125. Desaiah, D. and Koch, R. B., Inhibition of fish brain ATPases by aldrin-transidol, aldrin, dieldrin and photodieldrin, *Biochem. Biophys. Res. Commun.*, 64, 13, 1975.

126. Koch, R. B., Determination of the site(s) of action of selected pesticides by an enzymatic-immunobiological approach, EPA-600/3-78-093, U. S. Environmental Protection Agency, Cincinnati, 1978, 1.

127. Yap, H. H., Desaiah, D., Cutkomp, L. K., and Koch, R. B., *In vitro* inhibition of fish brain ATPase activity by cyclodiene insecticides and related compounds, *Bull. Environ. Contam. Toxicol.*, 14, 163, 1975.

128. Desaiah, D., Cutkomp, L. K., Vea, E. V., and Koch, R. B., The effect of three pyrethroids on ATPases of insects and fish, *Gen. Pharmacol.*, 6, 31, 1975.

129. Desaiah, D., Cutkomp, L. K., Koch, R. B., and Jarvinen, A., DDT: effect of continuous exposure on ATPase activity in fish, *Pimephales promelas*, *Arch. Environ. Contam. Toxicol.*, 3, 132, 1975.

130. Verma, S. R., Bansal, S. K., Gupta, A. K., and Dalela, R. C., *In vivo* effect on ATPase in certain tissues of *Labeo rohita* and *Saccobranchus fossilis*, following chronic chlordane intoxication, *Bull. Environ. Contam. Toxicol.*, 20, 769, 1978.

131. Jackson, D. A. and Gardner, D. R., The effects of some organochlorine pesticide analogs on salmonid brain ATPases, *Pestic. Biochem. Physiol.*, 2, 377, 1973.

132. Jackson, D. A. and Gardner, D. R., *In vitro* effects of DDT analogs on trout brain Mg^{2+}-ATPases. I. Specificity and physiological significance, *Pestic. Biochem. Physiol.*, 8, 113, 1978.

133. Jackson, D. A. and Gardner, D. R., *In vitro* effects of DDT analogs on trout brain Mg^{2+}-ATPases. II. Inhibition kinetics, *Pestic. Biochem. Physiol.*, 8, 123, 1978.

134. Desaiah, D. and Koch, R. B., Influence of solvents on the pesticide inhibition of ATPase activities in fish and insect tissue homogenates, *Bull. Environ. Contam. Toxicol.*, 17, 74, 1977.

135. Dalela, R. C., Bansal, S. K., Gupta, A. K., and Verma, S. R., Effect of solvents on *in vitro* pesticide inhibition of ATPase in certain tissues of *Labeo rohita, Water Air Soil Pollut.*, 11, 201, 1979.

136. Herzel, F. and Murty, A. S., Do carrier solvents enhance the water solubility of hydrophobic compounds?, *Bull. Environ. Contam. Toxicol.*, 32, 53, 1984.

137. Hylin, J. W., Pesticide residue analysis of water and sediments: potential problems and some philosophy, *Residue Rev.*, 76, 203, 1980.

138. *Standard Practice for Conducting Acute Toxicity Tests with Fishes, Macroinvertebrates, and Amphibians,* American Society for Testing and Materials, Philadelphia, 1980, 1.

139. Ewing, R. D., Johnson, S. L., Pribble, H. J., and Lichatowich, J. A., Temperature and photoperiod effects on gill (Na + K)-ATPase activity in chinook salmon (*Oncorhynchus tshawytscha*), *J. Fish. Res. Board Can.*, 36, 1347, 1979.

140. Zaugg, W. S., Advanced photoperiod and water temperature effects on gill Na^{+}-K^{+} adenosine triphosphate activity and migration of juvenile steelhead *(Salmo gairdneri), Can. J. Fish. Aquat. Sci.*, 38, 758, 1981.

141. Ewing, R. D., Pribble, H. J., Johnson, S. L., Fustish, C. A., Diamond, J., and Lichatowich, J. A., Influence of sex, growth rate, and photoperiod on cyclic changes in gill (Na + K)-ATPase activity in chinook salmon *(Oncorhynchus tshawytscha), Can. J. Fish. Aquat. Sci.*, 37, 600, 1980.

142. Sastry, K. V. and Sharma, S. K., The effect of *in vivo* exposure of endrin on the activities of acid, alkaline and glucose-6-phosphatases in liver and kidney of *Ophiocephalus (Channa) punctatus, Bull. Environ. Contam. Toxicol.*, 20, 456, 1978.

143. Sastry, K. V. and Sharma, S. K., *In vivo* effect of endrin on three phosphatases in kidney and liver of the fish *Ophiocephalus punctatus, Bull. Environ. Contam. Toxicol.*, 21, 185, 1979.

144. Sastry, K. V. and Sharma, S. K., Endrin toxicosis on few enzymes in liver and kidney of *Channa punctatus* (Bloch), *Bull. Environ. Contam. Toxicol.*, 22, 4, 1979.

145. Sharma, S. K. and Sastry, K. V., Alteration in enzyme activities in liver and kidney of *Channa punctatus* exposed to endrin, *Bull. Environ. Contam. Toxicol.*, 22, 17, 1979.

146. Sharma, S. K., Chaturvedi, L. D., and Sastry, K. V., Acute endrin toxicity on oxidases of *Ophiocephalus punctatus* (Bloch), *Bull. Environ. Contam. Toxicol.*, 23, 153, 1979.

147. Lane, C. E. and Scura, E. D., Effects of dieldrin on glutamic oxaloacetic transaminase in *Poecilia latipinna, J. Fish. Res. Board Can.*, 27, 1869, 1970.

148. Camp, B. J., Hejtmancik, E., Armour, C., and Lewis, D. H., Acute effects of Aroclor® 1254 (PCB) on *Ictalurus punctatus* (catfish), *Bull. Environ. Contam. Toxicol.*, 12, 204, 1974.

149. Siva Prasada Rao, K. and Ramana Rao, K. V., Regulation of phosphorylases and aldolases in tissues of the teleost (*Tilapia mossambica*) under methyl parathion impact, *Bull. Environ. Contam. Toxicol.*, 31, 474, 1983.

150. Siva Prasada Rao, K. and Ramana Rao, K. V., Effect of sublethal concentration of methyl parathion on selected oxidative enzymes and organic constitutents in the tissues of the freshwater fish, *Tilapia mossambica* (Peters), *Curr. Sci.*, 48, 526, 1979.

151. Siva Prasada Rao, K., Kabeer Ahammad, I., and Ramana Rao, K. V., Impact of methyl parathion on lactate dehydrogenase isozymes of a teleost *Tilapia mossambica (Peters), Indian J. Fish.*, 29, 185, 1982.

152. Verma, S. R., Tonk, I. P., Gupta, A. K., and Dalela, R. C., *In vivo* enzymatic alterations in certain tissues of *Saccobranchus fossilis* following exposure to four toxic substances, *Environ. Pollut.*, 26, 121, 1981.

153. McCorkle, F. M., Chambers, J. E., and Yarbrough, J. D., Tissue enzyme activities following exposure to dietary mirex in the channel catfish, *Ictalurus punctatus, environ. Pollut.*, 19, 195, 1979.

154. Tandon, R. S. and Dubey, A., Toxic effects of two organophosphorus pesticides on fructose-1, 6-diphosphate aldolase activity of liver, brain and gills of the freshwater fish *Clarias batrachus, Environ. Pollut.*, 31, 1, 1983.

155. Whitmore, D. H., Jr. and Hodges, D. H., Jr., *In vitro* pesticide inhibition of muscle esterases of the mosquitofish, *Gambusia affinis, Comp. Biochem. Physiol.*, 59C, 145, 1978.

156. Sastry, K. V. and Malik, P. V., Studies on the effect of dimecron on the digestive system of a freshwater fish, *Channa punctatus, Arch. Environ. Contam. Toxicol.*, 8, 397, 1979.

157. Sastry, K. V., Siddiqui, A. A., and Singh, S. K., Alteration in some biochemical and enzymological parameters in the snake head fish *Channa punctatus*, exposed chronically to quinolphos, *Chemosphere*, 11, 1211, 1982.

158. Sastry, K. V. and Malik, P. V., Acute and chronic effects of diazinon on the activities of three dehydrogenases in the digestive system of a freshwater teleost fish *Channa punctatus, Toxicol. Lett.,* 10, 55, 1982.

159. Sastry, K. V. and Malik, P. V., Histopathological and enzymological alterations in the digestive system of a freshwater teleost fish, *Heteropneustes fossilis,* exposed acutely and chronically to diazinon, *Ecotoxicol. Environ. Saf.,* 6, 223, 1982.

160. Dragomirescu, A., Raileanu, L., and Ababei, L., The effect of carbetox on glycolysis and the activity of some enzymes in carbohydrate metabolism in the fish and rat liver, *Water Res.,* 9, 205, 1975.

161. Verma, S. R., Rani, S., and Dalela, R. C., Effect of phenol on *in vivo* activity of tissue transaminases in the fish *Notopterus notopterus, Ecotoxicol. Environ. Saf.,* 6, 171, 1982.

162. Verma, S. R., Rani, S., and Dalela, R. C., Effects of phenol and dinitrophenol on acid and alkaline phosphatases in tissues of a fish (*Notopterus notopterus*), *Arch. Environ. Contam. Toxicol.,* 9, 451, 1980.

163. Dalela, R. C., Rani, S., and Verma, S. R., Physiological stress induced by sublethal concentrations of phenol and pentachlorophenol in *Notopterus notopterus:* hepatic acid and alkaline phosphatases and succinic dehydrogenase, *Environ. Pollut.,* 21, 3, 1980.

164. Bostrom, S. -L. and Johansson, R. G., Effects of pentachlorophenol on enzymes involved in energy metabolism in the liver of the eel, *Comp. Biochem. Physiol.,* 41, 359, 1972.

165. Schultz, D. P. and Harman, P. D., Effects of fishery chemicals on the *in vitro* activity of glucose-6-phosphate dehydrogenase, *Bull. Environ. Contam. Toxicol.,* 25, 203, 1980.

166. Pocker, Y., Beug, W. M., and Ainardi, V. R., Carbonic anhydrase interaction with DDT, DDE, and dieldrin, *Science,* 174, 1336, 1971.

167. Meany, J. E. and Pocker, Y., The *in vitro* inactivation of lactate dehydrogenase by organochlorine insecticides, *Pestic. Biochem. Physiol.,* 11, 232, 1979.

168. Leatherland, J. F., Sonstegard, R. A., and Holdrient, M. V., Effect of dietary mirex and PCB's on hepatosomic index, liver lipid, carcass lipid and PCB and mirex bioaccumulation in yearling coho salmon, *Oncorhynchus kisutch, Comp. Biochem. Physiol.,* 63C, 243, 1979.

169. Siva Prasada Rao, K. and Ramana Rao, K. V., Lipid derivatives in the tissues of the freshwater teleost, *Saurotherodon mossambicus* (alias *Tilapia mossambica*), *Proc. Indian Nat. Sci. Acad. Part B,* 47, 53, 1981.

170. Murty, A. S. and Priyamvada Devi, A., The effect of endosulfan and its isomers on tissue protein glycogen, and lipids in the fish *Channa punctata, Pestic. Biochem. Physiol.,* 17, 280, 1982.

171. Maki, A. W., Correlations between *Daphnia magna* and fathead minnow (*Pimephales promelas*) chronic toxicity values for several classes of test substances, *J. Fish. Res. Board Can.,* 36, 411, 1979.

172. Mehrle, P. M., Stalling, D. L., and Bloomfield, R. A., Serum amino acids in rainbow trout (*Salmo gairdneri*) as affected by DDT and dieldrin, *Comp. Biochem. Physiol.,* 388, 373, 1971.

173. Mehrle, P. M. and DeClue, M. E., Phenylalanine metabolism altered by dietary dieldrin, *Nature,* 238, 462, 1972.

174. Mukhopadhyay, P. K. and Dehadrai, P. K., Biochemical changes in the air-breathing catfish *Clarias batrachus* (Linn.) exposed to malathion, *Environ. Pollut.,* 22, 149, 1980.

175. Siva Prasada Rao, K., Sambasiva Rao, K. R. S., Sreenivasa Babu, K., and Ramana Rao, K. V., Toxic impact of phenthoate on some functional aspects of nitrogen metabolism in tissues of the fish *Channa punctatus, Indian J. Fish.,* 29, 229, 1982.

176. Kabeer Ahammad Sahib, I., Prasada Rao, K. S., Sambasiva Rao, K. R. S., and Ramana Rao, K. V., Sublethal toxicity of malathion on the proteases and free amino acid composition in the liver of the teleost, *Tilapia mossambica* (Peters), *Toxicol. Lett.,* 20, 59, 1984.

177. Stelzer, R., Rosenthal, H., und Siebers, D., EinfluB von 2,4-Dinitrophenol auf die Atmung und die Konzentration einger Metabolite bei Embryonen des Herings *Clupea harengus, Mar. Biol.,* 11, 369, 1971.

178. Singh, N. N. and Srivastava, A. K., Effects of endosulfan on fish carbohydrate metabolism, *Ecotoxicol. Environ. Saf.,* 5, 412, 1981.

179. Grant, B. F. and Mehrle, P. M., Endrin toxicosis in rainbow trout *(Salmo gairdneri), J. Fish. Res. Board Can.,* 30, 31, 1973.

180. Srivastava, A. K. and Singh, N. N., Effects of acute exposure to methyl parathion on carbohydrate metabolism of Indian catfish *(Heteropneustes fossilis), Acta Pharmacol. Toxicol.,* 48, 26, 1981.

181. Eisler, R. and Edmunds, P. H., Effects of endrin on blood and tissue chemistry of a marine fish, *Trans. Am. Fish. Soc.,* 95, 153, 1966.

182. Eisler, R., Tissue changes in puffers exposed to methoxychlor and methyl parathion, Technical papers No. 17, Fish and Wildlife Service, Bureau of Sport Fisheries and Wildlife, U.S. Department of the Interior, Washington, D. C., 1967, 1.

183. Leatherland, J. F. and Sonstegard, R. A., Effect of dietary mirex and PCBs on calcium and magnesium metabolism in rainbow trout, *Salmo gairdneri* and coho salmon, *Oncorhynchus kisutch;* a comparison with Great Lakes coho salmon, *Comp. Biochem. Physiol.,* 60, 345, 1981.

184. Kabeer Ahammad Sahib, I., Jagannatha Rao, K. S., and Ramana Rao, K. V., Effect of malathion exposure on some physical parameters of whole body and on tissue cations of teleost, *Tilapia mossambica* (Peters), *J. Biosci.,* 3, 17, 1981.

185. Siva Prasada Rao, K., Madhu, Ch., Sambasiva Rao, K. R. S., and Ramana Rao, K. V., Effect of methyl parathion on body weight, water content and ionic changes in the teleost, *Tilapia mossambica* (Peters), *J. Food Sci. Technol.,* 20, 27, 1983.

186. Seshagiri Rao, K., Sreenivasa Moorthy, K., Dhananjaya Naidu, M., Sreeramulu Chetty, C., and Swami, K. S., Changes in nitrogen metabolism in tissues of fish (*Sarotherodon mossambicus*) exposed to benthiocarb, *Bull. Environ. Contam. Toxicol.,* 30, 473, 1983.

187. Kabeer Ahammad Sahib, I., Prasada Rao, K. S., Madhu, Ch., and Ramana Rao, K. V., Effect of malathion on some functional aspects of nitrogen utility in the teleost, *Tilapia mossambica* (Peters), *Natl. Acad. Sci. Lett.,* 4, 417, 1981.

188. Siva Prasada Rao, K., Kabeer Ahammad, I., and Ramana Rao, K. V., Impact of methyl parathion on the tissue NH_3-changes in the fish *Tilapia mossambica* (Peters), *Proc. Indian Natl. Sci. Acad., Part B,* 47, 394, 1981.

189. Prein, A. E., Thie, G. M., Alink, G. M., and Koeman, J. H., Cytogenetic changes in fish exposed to water of the river Rhine, *Sci. Total Environ.,* 9, 287, 1978.

190. Barron, M. G. and Adelman, I. R., Nucleic acid, protein content and growth of larval fish sublethally exposed to various toxicants, *Can. J. Fish. Aquat. Sci.,* 41(1), 141, 1984.

191. Janicki, R. H., and Kinter, W. B., DDT: disrupted osmoregulatory events in the intestine of the eel *Anguilla rostrata* adapted to seawater, *Science,* 173, 1146, 1971.

192. Leadem, T. P., Campbell, R. D., and Johnson, D. W., Osmoregulatory responses to DDT and varying salinities in *Salmo gairdneri.* I. Gill Na-K-ATPase, *Comp. Biochem. Physiol.,* 49A, 197, 1974.

193. Waggoner, J. P., III and Zeeman, M. G., DDT: short term effects on osmoregulation in black surfperch (*Embiotoca jacksoni*), *Bull. Environ. Contam. Toxicol.,* 13, 297, 1975.

194. Neufeld, G. J. and Pritchard, J. B., An assessment of DDT toxicity on osmoregulation and gill NA,K-ATPase activity in the blue crab, in *Aquatic Toxicology, ASTM STP 667,* Marking, L. L. and Kimerle, R. A., Eds., American Society for Testing and Materials, Philadelphia, 1979, 23.

195. Mayer, F. L. and Mehrle, P. M., Toxicological aspects of toxaphene in fish: a summary, Trans. 42nd N. Am. Wildlife Resour. Conf. Wildlife Management Institute, Washington, D. C., 1977, 365.

196. Mehrle, P. M. and Mayer, F. L., Jr., Toxaphene effects on growth and bone composition of fathead minnows, *Pimephales promelas, J. Fish. Res. Board Can.,* 32, 593, 1975.

197. Mehrle, P. M. and Mayer, F. L., Jr., Toxaphene effects on growth and development of brook trout (*Salvelinus fontinalis*), *J. Fish. Res. Board Can.,* 32, 609, 1975.

198. Mehrle, P. M. and Mayer, F. L., Clinical tests in aquatic toxicology: state of the art, *Environ. Health Perspect.,* 34, 139, 1980.

199. Mayer, F. L., Mehrle, P. M., and Schoettger, R. A., Collagen metabolism in fish exposed to organic chemicals, in Recent Advances in Fish Toxicology, EPA-600/3-77-085, Tubb, R. A., Ed., U.S. Environmental Protection Agency, Corvallis, Ore., 1977, 31.

200. Hamilton, S. J., Mehrle, P. M., and Mayer, F. L., Mechanical properties of bone in channel catfish as affected by Vitamin C and toxaphene, *Trans. Am. Fish. Soc.,* 110, 718, 1981.

201. Moccia, R. D., Leatherland, J. F., and Sonstegard, R. A., Increasing frequency of thyroid goiters in coho salmon (*Oncorhynchus kisutch*) in the Great Lakes, *Science,* 198, 425, 1977.

202. Leatherland, J. F. and Sonstegard, R. A., Lowering of serum thyroxine and triiodothyronine levels in yearling coho salmon, *Oncorhynchus kisutch,* by dietary mirex and PCBs, *J. Fish. Res. Board Can.,* 35, 1285, 1978.

203. Leatherland, J. F. and Sonstegard, R. A., Effect of dietary mirex and PCB (Aroclor 1254) on thyroid activity and lipid reserves in rainbow trout *Salmo gairdneri* Richardson, *J. Fish. Dis.,* 2, 43, 1979.

204. Leatherland, J. F. and Sonstegard, R. A., Effect of dietary mirex and PCB's in combination with food deprivation and testosterone administration on thyroid activity and bioaccumulation of organochlorines in rainbow trout *Salmo gairdneri* Richardson, *J. Fish. Dis.,* 3, 115, 1980.

205. Leatherland, J. F. and Sonstegard, R. A., Bioaccumulation of organochlorines by yearling coho salmon (*Oncorhynchus kisutch* Walbaum) fed diets containing Great Lakes coho salmon, and the pathophysiological responses of the recipients, *Comp. Biochem. Physiol.,* 72C, 91, 1982.

206. Hilton, J. W., Hodson, P. V., Braun, H. E., Leatherland, J. F., and Slinger, S. J., Contaminant accumulation and physiological response in rainbow trout (*Salmo gairdneri*) reared on naturally contaminated diets, *Can. J. Fish. Aquat. Sci.,* 40, 1987, 1983.

207. Moccia, R. D., Leatherland, J. F., Sonstegard, R. A., and Holdrinet, M. V. H., Are goiter frequencies in Great Lakes salmon correlated with organochlorine residues?, *Chemospere*, 8, 649, 1978.

208. Leatherland, J. F. and Sonstegard, R. A., Thyroid responses in rats fed diets formulated with Great Lakes coho salmon, *Bull. Environ. Contam. Toxicol.*, 29, 341, 1982.

209. Freeman, H. C. and Sangalang, G., Changes in steroid hormone metabolism as a sensitive method of monitoring pollutants and contaminants, Proc. 3rd Aquatic Toxicity Workshop, EPS-5-AR-77-1, Environmental Protection Service, Halifax, Nova Scotia, Canada, 1977, 123.

210. Freeman, H. C. and Idler, D. H., The effect of polychlorinated biphenyl on steroidogenesis and reproduction in the brook trout *(Salvelinus fontinalis), Can. J. Biochem.*, 53, 666, 1975.

211. Freeman, H. C., Sangalang, G., and Flemming, B., The effects of a polychlorinated biphenyl (PCB) diet on Atlantic cod *(Gadus morhua), Int. Counc., Explor. Sea*, C. M. 1978/E:18, 1, 1978.

212. Freeman, H. C., Uthe, J. F., and Sangalang, G., The use of steroid hormone metabolism studies in assessing the sublethal effects of marine pollution, *Rapp. P. -V. Reun. Cons. Int. Explor. Mer.*, 179, 16, 1980.

213. Mayer, F. L., Mehrle, P. M., and Sanders, H. O., Residue dynamics and biological effects of polychlorinated biphenyls in aquatic organisms, *Arch. Environ. Contam. Toxicol.*, 5, 501, 1977.

214. Singh, H. and Singh, T. P., Short-term effect of two pesticides on the survival, ovarian ^{32}P uptake and gonadotrophic potency in a freshwater catfish, *Heteropneustes fossilis* (Bloch), *J. Endocrinol.*, 85, 193, 1980.

215. Singh, H. and Singh, T. P., Thyroid activity and TSH potency of the pituitary gland and blood serum in response to cythion and hexadrin treatment in the freshwater catfish, *Heteropneustes fossilis* (Bloch), *Environ. Res.*, 22, 184, 1980.

216. Singh, H. and Singh, T. P., Effect of two pesticides on ovarian ^{32}P uptake and gonadotrophin concentration during different phases of annual reproductive cycle in the freshwater catfish, *Heteropneustes fossilis* (Bloch), *Environ. Res.*, 22, 190, 1980.

217. Singh, H. and Singh, T. P., Effect of parathion and aldrin on survival, ovarian ^{32}P uptake and gonadotrophic potency in a freshwater catfish, *Heteropneustes fossilis* (Bloch), *Endokrinologie*, 77, 173, 1981.

218. Singh, H. and Singh, T. P., Effects of two pesticides on testicular ^{32}P uptake, gonadotrophic potency, lipid and cholesterol content of testis, liver and blood serum during spawning phase in *Heteropneustes fossilis* (Bloch), *Endokrinologie*, 76, 288, 1980.

219. Singh, H. and Singh, T. P., Effect of two pesticides on total lipid and cholesterol contents of ovary, liver and blood serum during different phases of the annual reproductive cycle in the freshwater teleost, *Heteropneustes fossilis* (Bloch), *Environ. Pollut.*, 23, 9, 1980.

220. Singh, H. and Singh, T. P., Effect of some pesticides on hypothalamo-hypophyseal-ovarian axis in the freshwater catfish *Heteropneustes fossilis* (Bloch), *Environ. Pollut.*, 27, 283, 1982.

221. Christensen, G. M., Fiandt, J. T., and Poeschl, B. A., Cells, proteins, and certain physical-chemical properties of brook trout *(Salvelinus fontinalis)* blood, *J. Fish. Biol.*, 12, 51, 1978.

222. Brungs, W. A. and Mount, D. I., Lethal endrin concentration in the blood of gizzard shad, *J. Fish. Res. Board Can.*, 24, 429, 1967.

223. Mount, D. I., Vigor, L. W., and Schafer, M. L., Endrin: use of concentration in blood to diagnose acute toxicity to fish, *Science*, 152, 1388, 1966.

224. Plumb, J. A. and Richburg, R. W., Pesticide levels in sera of moribund channel catfish from a continous winter mortality, *Trans. Am. Fish. Soc.*, 106, 185, 1977.

225. Witt, J. M., Brown, W. H., Shaw, G. I., Maynard, L. S., Sullivan, L. M., Whiting, F. M., and Stull, J. W., Rate of transfer of DDT from the blood compartment, *Bull. Environ. Contam. Toxicol.*, 1, 187, 1966.

226. Bridges, W. R., Kallman, J. B., and Andrews, K. A., Persistence of DDT and its metabolites in a farm pond, *Trans. Am. Fish. Soc.*, 92, 421, 1963.

227. Moss, J. A. and Hathway, D. E., Transport of organic compounds in the mammal: partition of dieldrin and telodrin between the cellular components and soluble proteins of blood, *Biochem. J.*, 91, 384, 1964.

228. Plack, P. A., Skinner, E. R., Rogie, A., and Mitchell, A. I., Distribution of DDT between the lipoproteins of trout serum, *Comp. Biochem. Physiol.*, 62C, 119, 1979.

229. Olson, K. R., Bergman, H. L., and Fromm, P. O., Uptake of methyl mercuric chloride and mercuric chloride by trout: a study of uptake pathways into the whole animal and uptake by erythrocytes *in vitro*, *J. Fish. Res. Board Can.*, 30, 1293, 1973.

230. Cope, O. B., Wood, E. M., and Wallen, G. H., Some chronic effects of 2,4-D on the bluegill *(Lepomis macrochirus)*, *Trans. Am. Fish. Soc.*, 99, 1, 1970.

231. Holmberg, B., Jensen, S., Larsson, A., Lewander, K., and Olsson, M., Metabolic effects of technical pentachlorophenol (PCP) on the eel *Anguilla anguilla* L, *Comp. Biochem. Physiol.*, 43B, 171, 1972.

232. Burton, A. and Sinsheimer, R. L., Insecticides: effects on cutthroat trout of repeated exposure to DDT, *Science*, 142, 958, 1963.

233. Johansson-Sjobeck, M.- L., Dave, G., and Lidman, U., Hematological effects of PCB (polychlorinated biphenyls) in the European eel, *Anguilla anguilla* L, and the rainbow trout, *Salmo gairdneri* Rich., in *Sublethal Effects of Toxic Chemicals on Aquatic Animals*, Koeman, J. H. and Strik, J. J. T. W. A., Eds., Elsevier, Amsterdam, 1975, 189.

234. McBride, J. R., Dye, H. M., and Donaldson, E. M., Stress response of juvenile sockeye salmon (*Oncorhynchus nerka*) to the butoxyethanol ester or 2,4-dichlorophenoxyacetic acid, *Bull. Environ. Contam. Toxicol.*, 27, 877, 1981.

235. Sastry, K. V. and Sharma, K., Diazinon-induced hematological changes in *Ophiocephalus* (*Channa*) *punctatus*, *Ecotoxicol. Environ. Saf.*, 5, 171, 1981.

236. Sastry, K. V. and Sharma, K., Diazinon-induced histopathological and hematological alterations in a freshwater teleost, *Ophiocephalus punctatus*, *Ecotoxicol. Environ. Saf.*, 5, 329, 1981.

237. Madder, D. J. and Lockhart, W. L., A preliminary study of the effects of diflubenzuron and methoprene on rainbow trout (*Salmo gairdneri* Richardson), *Bull. Environ. Contam. Toxicol.*, 20, 66, 1978.

238. Westman, I., Johansson-Sjobeck, M. -L., and Fange, R., The effect of PCB on the activity of delta-amino-levulinic acid dehydratase (ALA-D) and on some hematological parameters in the rainbow trout, *Salmo gairdneri*, in *Sublethal Effects of Toxic Chemicals on Aquatic Animals*, Koeman, J. H. and Strik, J. J. T. W. A., Eds., Elsevier, Amsterdam, 1975, 234.

239. Haux, C. and Larsson, A., Effects of DDT on blood plasma electrolytes in the flounder, *Platichthys flesus* L, in hypotonic brackish water, *Ambio*, 8, 171, 1979.

240. Uthe, J. F., Freeman, H. C., Mounib, S., and Lockhart, W. L., Selection of biochemical techniques for detection of environmentally induced sublethal effects in organisms, *Rapp. P. -V. Reun. Cons. Int. Explor. Mer.*, 179, 39, 1980.

241. Wedemeyer, G., Some physiological aspects of sublethal heat stress in the juvenile steelhead trout (*Salmo gairdneri*) and coho salmon (*Oncorhynchus kisutch*), *J. Fish. Res. Board Can.*, 30, 831, 1973.

242. Matthiessen, P., Haematological changes in fish following aerial spraying with endosulfan insecticide for tsetse fly control in Botswana, *J. Fish. Biol.*, 18, 461, 1981.

243. Stevens, E. D., Change in body weight caused by handling and exercise in fish, *J. Fish. Res. Board Can.*, 29, 202, 1972.

244. Wedemeyer, G., Stress of anesthesia with M. S. 222 and benzocaine in rainbow trout (*Salmo gairdneri*), *J. Fish. Res. Board Can.*, 27, 909, 1970.

245. O'Connor, D. V. and Fromm, P. O., The effect of methyl mercury on gill metabolism and blood paramters of rainbow trout, *Bull. Environ. Contam. Toxicol.*, 13, 406, 1975.

246. Weil, C. S. and Carpenter, C. P., Abnormal values in control groups during repeated-dose toxicologic studies, *Toxicol. Appl. Pharmacol.*, 14, 335, 1969.

247. Eller, L. L., Histopathologic lesions in cutthroat trout (*Salmo clarki*) exposed chronically to the insecticide endrin, *Am. J. Pathol.*, 64, 321, 1971.

248. Couch, J. A., Histopathological effects of pesticide and related chemicals on the livers of fishes, in *The Pathology of Fishes*, William, E. R. and Migoki, G., Eds., University of Wisconsin Press, Madison, 1975, 559.

249. Lowe, J. I., Chronic exposure of spot, *Leiostomus xanthurus*, to sublethal concentrations of toxaphene in sea water, *Trans. Am. Fish. Soc.*, 93, 396, 1964.

250. Weis, P., Ultrastructural changes induced by low concentrations of DDT in the livers of the zebrafish and guppy, *Chem. Biol. Interact.*, 8, 25, 1974.

251. Sastry, K. V. and Sharma, S. K., The effect of endrin on the histopathological changes in the liver of *Channa punctatus*, *Bull. Environ. Contam. Toxicol.*, 20, 674, 1978.

252. Sastry, K. V. and Sharma, S. K., Endrin toxicity on liver of *Channa punctatus* (Bloch), *Indian J. Exp. Biol.*, 16, 372, 1978.

253. Sastry, K. V. and Sharma, S. K., Endrin induced hepatic injury in *Channa punctatus* (Ham.), *Indian. J. Fish.*, 26, 250, 1979.

254. VanValin, C. C., Andrews, A. K., and Eller, L. L., Some effects of mirex on two warm water fishes, *Trans. Am. Fish. Soc.*, 97, 185, 1968.

255. Goodman, L. R., Hansen, D. J., Couch, J. A., and Forester, J., Effects of heptachlor and toxaphene on laboratory-reared embryos and fry of the sheepshead minnow, *Southeastern Assoc. Game Fish Comm. 30th Annu. Conf.*, 1976, 192.

256. Parrish, P. R., Couch, J. A., Forester, J., Patrick, J. M., Jr., and Cook, G. H., Dieldrin: effects on several estuarine organisms, *27th Annu. Conf. Southeastern Assoc. Game Fish Comm.*, 1973, 427.

257. Lipsky, M. M., Klauning, J. E., and Hinton, D. E., Comparison of acute response to polychlorinated biphenyl in liver of rat and channel catfish: a biochemical and morphological study, *J. Toxicol. Environ. Health*, 4, 107, 1978.

258. Nimmo, D. R., Hansen, D. J., Couch, J. A., Cooley, N. R., Parrish, P. R., and Lowe, J. I., Toxicity of Aroclor® 1254 and its physiological activity in several estuarine organisms, *Arch. Environ. Contam. Toxicol.*, 3, 22, 1974.

259. Sangalang, G. B., Freeman, H. C., and Crowell, R., Testicular abnormalities in cod (*Gadus morhua*) fed Aroclor 1254, *Arch. Environ. Contam. Toxicol.*, 10, 617, 1981.

260. Hendricks, J. D., Putnam, T. P., Bills, D. D., and Sinnhuber, R. O., Inhibitory effect of a polychlorinated biphenyl (Aroclor 1254) on aflatoxin B₁ carcinogenesis in rainbow trout (*Salmo gairdneri), J. Natl. Cancer Inst.*, 59, 1545, 1977.

261. Kumaraguru, A. K., Beamish, F. W. H., and Ferguson, H., W., Direct and circulatory paths of permethrin (NRDC-143) causing histopathological changes in the gills of rainbow trout, *Salmo gairdneri* Richardson, *J. Fish. Biol.*, 20, 87, 1982.

262. Kendall, M. W., Acute effects of methyl mercury toxicity in channel catfish (*Ictalurus punctatus*) kidney, *Bull. Environ. Contam. Toxicol.*, 13, 570, 1975.

263. Kendall, M. W., Acute effects of methyl mercury toxicity in channel catfish (*Ictalurus punctatus*) liver, *Bull. Environ. Contam. Toxicol.*, 18, 143, 1977.

264. Anees, M. A., Hepatic pathology in a freshwater teleost *Channa punctatus* (Bloch) exposed to sublethal and chronic levels of three organophosphorus insecticides, *Bull. Environ. Contam. Toxicol.*, 19, 524, 1978.

265. Sprague, J. B., Measurement of pollutant toxicity to fish. III. Sublethal effects and "SAFE" concentrations, *Water Res.*, 5, 245, 1971.

266. Owen, J. W. and Rosso, S. W., Effects of sublethal concentrations of phentachlorophenol on the liver of bluegill sunfish, *Lepomis macrochirus, Bull. Environ. Contam. Toxicol.*, 26, 594, 1981.

267. Lingaraja, T. and Venugopalan, V. K., Pesticide-induced physiological and behavioural changes in an estuarine teleost *Therapon jarbua* (Forsk), *Fish. Technol.*, 15, 115, 1978.

268. Verma, S. R., Bansal, S. K., and Dalela, R. C., Toxicity of selected organic pesticides to a freshwater teleost fish, *Saccobranchus fossilis* and its application in controlling water pollution, *Arch. Environ. Contam. Toxicol.*, 7, 317, 1978.

269. Waiwood, K. G. and Johansen, P. H., Oxygen consumption and activity of the white sucker (*Catostomus commersoni),* in lethal and nonlethal levels of the organochlorine insecticide, methoxychlor, *Water Res.*, 8, 401, 1974.

270. Rodger, D. W. and Beamish, F. W. H., Uptake of waterborne methylmercury by rainbow trout (*Salmo gairdneri)* in relation to oxygen consumption and methylmercury concentration, *Can. J. Fish. Aquat. Sci.*, 38, 1309, 1981.

271. Rao, D. M. R., Devi, A. P., and Murty, A. S., Toxicity and metabolism of endosulfan and its effect on oxygen consumption and total nitrogen excretion of the fish *Macrognathus aculeatum, Pestic. Biochem. Physiol.*, 15, 282, 1981.

272. Rao, D. M. R., Devi, A. P., and Murty, A. S., Relative toxicity of endosulfan, its isomers, and formulated products to the freshwater fish *Labeo rohita, J. Toxicol. Environ. Health*, 6, 825, 1980.

273. Murty, A. S., Rajabhushanam, B. R., Ramani, A. V., and Christopher, K., Toxicity of fenitrothion to the fish *Mystus cavasius* and *Labeo rohita, Environ. Pollut.*, 30, 225, 1983.

274. Crandall, C. A. and Goodnight, C. J., Effects of sublethal concentrations of several toxicants on growth of the common guppy, *Lebistes reticulatus, Linnol. Oceanogr.*, 7, 233, 1962.

275. Kumaraguru, A. K. and Beamish, F. W. H., Bioenergetics of acclimation to permethrin (NRDC-143) by rainbow trout, *Comp. Biochem. Physiol.*, 75, 247, 1983.

276. Rath, S. and Misra, B. N., Age-related changes in oxygen consumption by the gill, brain and muscle tissues of *Tilapia mossambica* Peters exposed to dichlorvos (DDVP), *Environ. Pollut.*, 23, 95, 1980.

277. Hiltibran, R. C., Oxygen and phosphate metabolism of bluegill liver mitochondria in the presence of some pesticides, in *Environmental Quality and Safety, Suppl. III.*, Conlston, F. and Corte, F., Eds., Georg Thieme, Stuttgart, 1975, 880.

278. O'Hara, J., A continuously monitored respiration chamber for fish, *Water Res.*, 5, 143, 1971.

279. Heath, A. G., A critical comparison of methods for measuring fish respiratory movements, *Water Res.*, 6, 1, 1972.

280. Morgan, W. S. G. and Kuhn, P. C., A method to monitor the effects of toxicants upon breathing rate of largemouth bass (*Micropterus salmoides* Lacépède), *Water Res.*, 8, 67, 1974.

281. Drummond, R. A., Disposable electrode chamber for measuring opercular movements of fathead minnows, *Prog. Fish-Culturist*, 38, 94, 1976.

282. Maki, A. W., Stewart, W., and Silvey, J. K. G., The effects of dibrom on respiratory activity of the stonefly, *Hydroperla crosbyi*, hellgrammite, *Corydalus cornutus* and the golden shiner, *Notemigonus crysoleucas, Trans. Am. Fish. Soc.*, 102, 806, 1973.

283. Bardach, J. E., Fujiya, A., and Holl, A., Detergents: effects on the chemical senses of the fish *Ictalurus natalis* (Le Saeur), *Science*, 148, 1605, 1965.

284. Cairns, J., Jr. and Loos, J. J., Changed feeding rate of *Brachydanio rario* (Hamilton Buchanan) resulting from exposure to sublethal concentrations of zinc, potassium dichromate, and alkyl benzene sulfonate detergent, *Proc. Pa. Acad. Sci.*, 11, 47, 1967.

285. Buhler, D. R. and Shanks, W. E., Influence of body weight on chronic oral DDT toxicity in coho salmon, *J. Fish. Res. Board Can.*, 27, 347, 1970.

286. Kleerekoper, H., Effects of exposure to a subacute concentration of parathion on the interaction between chemoreception and water flow in fish, in *Pollution and Physiology of Marine Organisms*, Vernberg, F. and Vernberg, W. B., Eds., Academic Press, New York, 1974, 237.

287. Arunachalam, S., Jeyalakshmi, K., and Aboobucker, S., Toxic and sublethal effects of carbaryl on a freshwater catfish, *Mystus vittatus* (Bloch), *Arch. Environ. Contam. Toxicol.*, 9, 307, 1980.

288. Folmar, L. C. and Hodgins, H. O., Effects of Aroclor 1254 and no. 2 fuel oil, singly and in combination, on predator-prey interactions in coho salmon (*Oncorhynchus kisutch*), *Bull. Environ. Contam. Toxicol.*, 29, 24, 1982.

289. Tagatz, M. E., Effect of mirex on predator-prey interaction in an experimental estuarine ecosystem, *Trans. Am. Fish. Soc.*, 105, 546, 1976.

290. Macek, K. J., Reproduction in brook trout (*Salvelinus fontinalis*) fed sublethal concentrations of DDT, *J. Fish. Res. Board Can.*, 25, 1787, 1968.

291. Macek, K. J., Growth and resistance to stress in brook trout fed sublethal levels of DDT, *J. Fish. Res. Board Can.*, 25, 2443, 1968.

292. Buck, D. H., Baur, R. J., and Rose, C. R., Comparison of the effects of grass crap and the herbicide diuron in densely vegetated pools containing golden shiners and bluegills, *Prog. Fish-Culturist*, 37, 185, 1975.

293. Bengtsson, B. -E., Increased growth in minnows exposed to PCBs, *Ambio*, 8, 169, 1979.

294. Seelye, J. G. and Mac, M. J., Size-specific mortality in fry of lake trout (*Salvelinus namaycush*) from Lake Michigan, *Bull. Environ. Contam. Toxicol.*, 27, 376, 1981.

295. Niimi, A. J. and McFadden, C. A., Uptake of sodium pentachlorophenate (NaPCP) from water by rainbow trout (*Salmo gairdneri*) exposed to concentrations in the ng/L range, *Bull. Environ. Contam. Toxicol.*, 28, 11, 1982.

296. Webb, P. W. and Brett, J. R., Effects of sublethal concentrations of sodium pentachlorophenate on growth rate, food conversion efficiency, and swimming performance in underyearling sockeye salmon (*Oncorhynchus nerka*), *J. Fish. Res. Board Can.*, 30, 499, 1973.

297. Mayer, F. L., Jr., Mehrle, P. M., Jr., and Dwyer, W. P., Toxaphene effects on reproduction, growth, and mortality of brook trout, EPA-600/3-75-013, U.S. Environmental Protection Agency, Duluth, Minn., 1975, 1.

298. Mauck, W. L., Mehrle, P. M., and Mayer, F. L., Effects of the polychlorinated biphenyl Aroclor® 1254 on growth, survival, and bone development in brook trout (*Salvelinus fontinalis*), *J. Fish. Res. Board Can.*, 35, 1084, 1978.

299. Woodward, D. F., Toxicity of the herbicides dinoseb and picloram to cutthroat (*Salmo clarki*) and lake trout (*Salvelinus namaycush*), *J. Fish. Res. Board Can.*, 33, 1671, 1976.

300. Woodward, D. F., Assessing the hazard of picloram to cutthroat trout, *J. Range Manage.*, 32, 230, 1979.

301. Hermanutz, R. O., Endrin and malathion toxicity to flagfish (*Jordanella floridae*), *Arch. Environ. Contam. Toxicol.*, 7, 159, 1978.

302. Hansen, I. G., Wiekhorst, W. B., and Simon, J., Effects of dietary Aroclor® 1242 on channel catfish (*Ictalurus punctatus*) and the selective accumulation of PCB components, *J. Fish. Res. Board Can.*, 33, 1343, 1976.

303. Wildish, D. J. and Lister, N. A., Effects of dietary fenitrothion on growth and hierarchical position in brook trout, *Prog. Fish-Culturist*, 39, 3, 1977.

304. Sodergren, A., Dijirsarai, R., Gharibzadeh, M., and Moinpour, A., Organochlorine residues in aquatic environments in Iran, 1974, *Pestic. Monit. J.*, 12, 81, 1978.

305. Grzenda, A. R., Taylor, W. J., and Paris, D. F., The elimination and turnover of ^{14}C-dieldrin by different goldfish tissues, *Trans. Am. Fish. Soc.*, 101, 686, 1972.

306. Klaassen, H. E. and Kadoum, A. M., Pesticide residues in natural fish populations of the Smoky Hill River of Western Kansas-1967-69, *Pestic. Monit. J.*, 7, 53, 1973.

307. Bulkley, R. V., Kellogg, R. L., and Shannon, L. R., Size-related factors associated with dieldrin concentrations in muscle tissue of channel catfish *Ictalurus punctatus*, *Trans. Am. Fish. Soc.*, 105, 301, 1976.

308. Guiney, P. D., Melancon, M. J., Jr., Lech, J. J., and Peterson, R. E., Effects of egg and sperm maturation and spawning on the distribution and elimination of a polychlorinated biphenyl in rainbow trout (*Salmo gairdneri*), *Toxicol. Appl. Pharmacol.*, 47, 261, 1979.

309. Hose, J. E., Hannah, J. B., Landolt, M. L., Miller, B. S., Felton, S. P., and Iwaoka, W. T., Uptake of benzo(a)pyrene by gonadal tissue of flatfish (family Pleuronectidae) and its effects on subsequent egg development, *J. Toxicol. Environ. Health*, 7, 991, 1981.

310. Godsi, P. J. and Johnson, W. C., Pesticide monitoring of the aquatic biota at the Tule Lake National Wildlife Refuge, *Pestic. Monit. J.*, 1, 21, 1968.

311. Cuerrier, J.-P., Keith, J. A., and Stone, E., Problems with DDT in fish culture operations, *Nat. Can.*, 94, 315, 1967.

312. Holden, A. V., Organochlorine insecticide residues in salmonid fish, *J. Appl. Ecol.*, 3, 45, 1966.

313. Crawford, R. B. and Guarino, A. M., Effects of DDT in *Fundulus:* studies on toxicity, fate, and reproduction, *Arch. Environ. Contam. Toxicol.*, 4, 334, 1976.

314. Hansen, D. J., Willson, A. J., Nimmo, D. W. R., Schimmel, S. C., Bahn, L. L., and Huggett, R., Kepone: hazard to aquatic organisms, *Science*, 193, 528, 1976.

315. Butler, P. A., Pesticides in the estuary, *Proc. Marsh Estuary Manage. Symp.*, Newsom, J. D., Ed., Thos J. Moran's Sons, Baton Rouge, La., 1968, 252.

316. Boyd, C. E., Insecticides cause mosquitofish to abort, *Prog. Fish-Culturist*, 26, 138, 1963.

317. Brooker, M. P. and Edwards, R. W., Aquatic herbicides and the control of water weeds, *Water Res.*, 9, 1, 1975.

318. McIntyer, J. D., Toxicity of methyl mercury to steelhead trout sperm, *Bull. Environ. Contam. Toxicol.*, 9, 98, 1973.

319. Smith, R. K. and Cole, C. F., Effects of egg concentrations of DDT and dieldrin on development in winter flounder *(Pseudopleuronectes americanus)*, *J. Fish. Res. Board Can.*, 30, 1894, 1973.

320. Stock, J. N. and Cope, O. B., Some effects of TEPA, an insect chemosterilant on the guppy, *Poecilia reticulata*, *Trans. Am. Fish. Soc.*, 98, 280, 1969.

321. Kapur, K., Kampaldeep, K., and Toor, H. S., The effect of fenitrothion on reproduction of the teleost fish *Cyprinus carpio communis* Linn: a biochemical study, *Bull. Environ. Contam. Toxicol.*, 20, 438, 1978.

322. Curtis, L. R. and Beyers, R. J., Inhibition of oviposition in the teleost *Oryzias latipes*, induced by subacute Kepone exposure, *Comp. Biochem. Physiol.*, 61, 15, 1978.

323. Carlson, A. R., Effects of long-term exposure to carbaryl (Sevin) on minnow (*Pimephales promelas*), *J. Fish. Res. Board Can.*, 29, 583, 1971.

324. Yasuno, M., Hatakeyama, S., and Miyashita, M., Effects on reproduction in the guppy (*Poecilia reticulata*) under chronic exposure to temephos and fenitrothion, *Bull. Environ. Contam. Toxicol.*, 25, 29, 1980.

325. Holden, A. V., The effects of pesticides on life in freshwaters, *Proc. R. Soc. London, Ser. B.*, 180, 383, 1972.

326. Wilbur, R. L. and Whitney, E. W., Toxicity of the herbicide Kuron® (Silvex) to bluegill eggs and fry, *Trans. Am. Fish. Soc.*, 102, 630, 1973.

327. Halter, M. T. and Johnson, H. E., Acute toxicities of a polychlorinated biphenyl (PCB) and DDT alone and in combination to early life stages of coho salmon (*Oncorhynchus kisutch*), *J. Fish. Res. Board Can.*, 31, 1543, 1974.

328. Weis, J. S. and Weis, P. S., Optical malformations induced by insecticides in embryos of the Atlantic silverside, *Menidia menidia*, *Fish. Bull.*, 74, 208, 1976.

329. Solomon, H. M. and Weis, J. S., Abnormal circulatory development in medaka caused by the insecticides carbaryl, malathion and parathion, *Teratology*, 19, 51, 1979.

330. Venugopalan, V. K. and Sasibhushana Rao, P., Pesticide induced impairment in incubation and post embryonic development of planktonic eggs of estuarine fish of Porto Novo (South India) waters, Proc. Symp. Environ. Biol., Academy of Environmental Biology, Muzaffarnagar, India, 397, 1979.

331. Stauffer, T. M., Effects of DDT and PCB's on survival of lake trout eggs and fry in a hatchery and in Lake Michigan, 1973—1976, *Trans. Am. Fish. Soc.*, 108, 178, 1979.

332. Westernhagen, H. V., Rosenthal, H., Dethlefsen, V., Ernst, W., Harms, U., and Hansen, P. -D., Bioaccumulating substances and reproductive success in Baltic flounder *Platichthys flesus*, *Aquat. Toxicol.*, 1, 85, 1981.

333. Burdick, G. E., Harris, E. J., Dean, H. J., Walker, T. M., Skea, J., and Colby, D., The accumulation of DDT in lake trout and the effect on reproduction, *Trans. Am. Fish. Soc.*, 93, 127, 1964.

334. Hiltibran, R. C., Effects of some herbicides on fertilized fish eggs and fry, *Trans. Am. Fish. Soc.*, 96, 414, 1967.

335. Schimmel, S. C., Parrish, P. R., Hansen, D. J., Patrick, J. M., Jr., and Forester, J., Endrin: effects on several estuarine organisms, Proc. 28th Annu. Conf. Southeastern Assoc. Game Fish Comm., 1974, 187.

336. Hogan, J. W. and Brauhn, J. L., Abnormal rainbow trout fry from eggs containing high residues of a PCB (Aroclor 1242), *Prog. Fish-Culturist*, 37, 229, 1975.

337. Virtanen, M. T. and Hattula, M. L., The fate of 2,4,6-trichlorophenol in an aquatic continuous-flow system, *Chemosphere*, 11, 641, 1982.

338. Weis, P. and Weis, J. S., Toxicity of the PCBs Aroclor 1254 and 1242 to embryos and larvae of the mummichog, *Fundulus heteroclitus, Bull. Environ. Contam. Toxicol.*, 28, 298, 1982.

339. Cairns, J., Jr. and Dickson, K. L., Field and laboratory protocols for evaluating the effects of chemical substances on aquatic life, *J. Test. Eval.*, 6, 81, 1978.

340. Penrose, D. L. and Lenat, D. R., Effects of apple orchard runoff on the aquatic macrofauna of a mountain stream, *Arch. Environ. Contam. Toxicol.*, 11, 383, 1982.

341. Butler, P. A., Pesticides in the marine environment, *J. Appl. Ecol.*, 3, 253, 1966.

342. Hannon, M. R., Greichus, Y. A., Applegate, R. L., and Fox, A. C., Ecological distribution of pesticides in Lake Poinsett, South Dakota, *Trans. Am. Fish. Soc.*, 99, 496, 1970.

343. U.S. EPA Ambient water quality criteria for DDT, EPA-440/5-80-038, U.S. Environmental Protection Agency, Washington, D.C., 1980.

344. Welch, E. B. and Spindler, J. C., DDT persistence and its effect on aquatic insects and fish after an aerial application, *J. Water Pollut. Cont. Fed.*, 36, 1285, 1964.

345. Wallace, R. R. and Hynes, H. B. N., The castastrophic drift of stream insects after treatments with methoxychlor (1,1,1-trichloro-2,2-bis(p-methoxyphenyl) ethane), *Environ. Pollut.*, 8, 255, 1975.

346. Wallace, R. R., Hynes, H. B. N., and Merritt, W. F., Laboratory and field experiments with methoxychlor as a larvicide for Simulidae (Diptera), *Environ. Pollut.*, 10, 251, 1976.

347. Haines, T. A., Effect of an aerial application of carbaryl on brook trout (*Salvelinus fontinalis*), *Bull. Environ. Contam. Toxicol.*, 27, 534, 1981.

348. Elson, P. F., Meister, A. L., Saunders, J. W., Saunders, R. L., Sprague, J. B., and Zitko, V., Impact of chemical pollution on Atlantic Salmon in North America, Int. Atlantic Salmon Symp. 1973, International Atlantic Salmon Foundation, St. Andrews, New Brunswick, Canada.

349. Keenleyside, M. H. A., Effects of forest spraying with DDT in New Brunswick on food of young Atlantic salmon, *J. Fish. Res. Board Can.*, 24, 807, 1967.

350. Ide, F. P., Effects of forest spraying with DDT on aquatic insects of salmon streams in New Brunswick, *J. Fish. Res. Board Can.*, 24, 769, 1967.

351. Smies, M., Evers, R. H. J., Peijnenburg, F. H. M., and Koeman, J. H., Environmental aspects of field trails with pyrethroids to eradicate tsetse fly in Nigeria, *Ecotoxicol. Environ. Saf.*, 4, 114, 1980.

352. Bays, L. R., Pesticide pollution and the effects on the biota of Chew Valley Lake, *Environ. Pollut.*, 1, 205, 1971.

353. Hurlbert, S. H., Mulla, M. S., Keith, J. O., Westlake, W. E., and Dusch, M. E., Biological effects and persistence of Dursban® in freshwater ponds, *J. Econ. Entomol.*, 63, 43, 1970.

354. Nicholson, H. P., Webb, H. J., Lauer, G. J., O'Brien, R. E., Grzenda, A. R., and Shanklin, D. W., Insecticide contamination in a farm pond. I. Origin and duration, *Trans. Am. Fish. Soc.*, 91, 213, 1962.

355. Grzenda, A. R., Lauer, G. J., and Nicholson, H. P., Insecticide contamination in a farm pond. II. Biological effects, *Trans. Am. Fish. Soc.*, 91, 213, 1962.

356. Colwell, A. E. and Schaefer, C. H., Diets of *Ictalurus nebulosus* and *Pomoxis nigromaculatus* altered by diflubenzuron, *Can. J. Fish. Aquat. Sci.*, 37, 632, 1980.

357. Johnson, W. W. and Finley, M. T., Handbook of Acute Toxicity of Chemicals to Fish and Aquatic Invertebrates, U.S. Department of the Interior, Fish and Wildlife Service, Washington, D. C., 1980.

358. Hughes, D. N., Boyer, M. G., Papst, M. H., Fowle, C. D., Rees, G. A. V., and Baulu, P., Persistence of three organophosphorus insecticides in artificial ponds and some biological implications, *Arch. Environ. Contam. Toxicol.*, 9, 269, 1980.

359. Crosby, D. G., and Tucker, R. K., Toxicity of aquatic herbicides to *Daphnia magna, Science*, 154, 289, 1966.

360. George, J. P., Hingorani, H. G., and Rao, K. S., Herbicide toxicity to fish-food organisms, *Environ. Pollut.*, 28, 183, 1982.

361. Buchanan, D. V., Mollemann, R. E., and Stewart, N. E., Effects of the insecticide Sevin on various stages of the Dungeness crab, *Cancer magister, J. Fish. Res. Board Can.*, 27, 93, 1970.

362. Fremling, C. R. and Mauck, W. L., Methods for using nymphs of burrowing mayflies (Ephemeroptera, Hexagenia) as toxicity test organisms, in *Aquatic Invertebrate Bioassays*, ASTM STP 715, Buikema, A. L., Jr. and Cairns, J., Jr., Eds., American Society for Testing and Materials, Philadelphia, 1980, 81.

363. Harper, D. B., Smith, R. V., and Gotto, D. M., BHC residues of domestic orgin: a significant factor in pollution of freshwater in Northern Ireland, *Environ. Pollut.*, 12, 223, 1977.

364. McLeese, D. W., Metcalfe, C. D., and Zitko, V., Lethality of permethrin, cypermethrin and fenvalerate to salmon, lobster and shrimp, *Bull. Environ. Contam. Toxicol.*, 25, 950, 1980.

365. Gaufin, A. R., Jensen, L., and Nelson, T., Bioassays determine pesticide toxicity to aquatic invertebrates, *Water Sewage Works,* 108, 355, 1961.

366. Mosser, J. L., Fisher, N. S., and Wurster, C. F., Polychlorinated biphenyls and DDT alter species composition in mixed cultures of algae, *Science,* 176, 533, 1972.

367. Tagatz, M. E., Ivey, J. M. Moore, J. C., and Tobia, M., Effects of pentachlorophenol on the development of estuarine communities, *J. Toxicol. Environ. Health,* 3, 506, 1977.

368. Hansen, S. R. and Garton, R. R., The effects of diflubenzuron on a complex laboratory stream community, *Arch. Environ. Contam. Toxicol.,* 11, 1, 1982.

369. Schauerte, W., Lay, J. P., Klein, W., and Korte, F., Influence of 2,4,6-trichlorophenol and pentachlorophenol on the biota of aquatic systems, *Chemosphere,* 11, 71, 1982.

370. Tagatz, M. E., Ivey, J. M., Lehman, H. K., and Oglesby, J. L., Effects of Sevin in development of experimental estuarine communities, *J. Toxicol. Environ. Health,* 5, 643, 1979.

371. Heckman, C. W., Long-term effects of intensive pesticide applications on the aquatic community in orchard drainage ditches near Hamburg, Germany, *Arch. Environ. Contam. Toxicol.,* 10, 393, 1981.

372. Swingle, H. S., Fish populations in Alabama rivers and impoundments, *Trans. Am. Fish. Soc.,* 83, 47, 1954.

373. Brungs, W. A. and Mount, D. I., Introduction to a discussion of the use of aquatic toxicity tests for evaluation of the effects of toxic substances, in *Estimating the Hazard of Chemical Substances to Aquatic Life,* ASTM STP 657, Cairns, J., Jr., Dickson, K. L., and Maki, A. W., Eds., American Society for Testing and Materials, Philadelphia, 1978, 15.

Chapter 7

POLYCHLORINATED BIPHENYLS AND RELATED COMPOUNDS

I. INTRODUCTION

Soon after the electron capture gas chromatography was developed as a sensitive analytical tool, the occurrence of many gas chromatographic peaks with almost the same retention times as those of the DDT-group compounds was reported.[1,2] Although sometimes it was noted that these additional peaks did not correspond exactly with those of DDT or its metabolites,[3] they were thought to represent either the DDT group or the metabolites of many common organochlorine (OC) compounds. Their true nature remained unknown until 1966 when Jensen[4] correctly identified them as belonging to the polychlorinated biphenyls (PCBs).

PCBs were first synthesized in 1881, but their large-scale use in industry did not begin until 1930. For about 4 decades thereafter, they were the most commonly used industrial chemicals, with a wide range of applications. Their most common use was as coolant and insulation fluids in transformers. Besides, they were used as plasticizers, high-pressure hydraulic fluids, lubricants, cutting oils, heat transfer agents, and adhesives; they were also used for impregnating cotton and asbestos for purposes of insulation, in epoxy paints in protective coating of wood and metals, ballasts for fluorescent lamps, carbonless copying paper, etc. PCBs were marketed under different names (Aroclors®, Phenochlor, Clophen, Kenneclor, Pyralene, Decachlorodiphenyl, Enchlor®, Santotherm®, Therminal FR-1, etc). and they have completely permeated the environment, although they were never deliberately sprayed or used in the open. In 45 years of use (up to 1978), 635 million kg or PCBs were produced in the U.S. alone, of which one third reside in landfills, air and water, soil, and sediments.[5] Hence, it is not surprising to find PCB quantities as high as 540 to 2980 mg/kg in sediments (in Hudson River, just south of the General Electric plant near Fort Edward, N.Y.[6] and 133 to 213 mg/kg in fish (in 1970),[7] and to note that 91% of U.S. residents have detectable PCB residues in their tissues.[8]

Although instances of skin disorders in persons occupationally exposed to PCBs were known, until 1966 the possible environmental hazards of PCBs were little appreciated. Disasters such as the "Yusho episode" in Japan, resulting from the leakage of PCBs from faulty heat exchangers into rice oil, helped focus attention on PCBs,[9] along with the efforts of environmental toxicologists interested in quantifying pesticide residues in environmental samples. It is really difficult to assess whether PCBs have ever been used in combination with pesticides. Reynolds[10] drew attention to a Monsanto Co. technical bulletin that stated that because "Aroclors® in formulations trap and hold more volatile ingredients, they make volatile insecticides and repellents 'last longer' in residual activity." He also cited earlier experimental works where PCBs were used to lower the vapor pressure of insecticides with a view to enhance the kill-life of the latter. PCB-containing waste oils were used in Canada to dedust rural roads and as a diluent for herbicide spraying.[11] Despite the fact that PCBs were rarely, if ever, used in combination with pesticides, PCBs are of interest to environmental toxicologists studying pesticide residues for two reasons. First, the chemistry and environmental behavior of PCBs are similar to those of DDT. PCBs are recalcitrant compounds that are thermally stable and resistant to both acidic and basic hydrolysis. Further, they are nonoxidizing, inert, and resistant to the action of corrosives. They have low vapor pressure and are poorly soluble in water, highly nonpolar, and hence lipophilic. Second, because of their nonpolar nature, they are co-extracted and co-eluted with OC pesticides when routine analytical methods are employed. Thus, unless care is taken to separate PCBs from OC compounds, at least 20 of the possible 209 PCB isomers appear together with OC compounds in packed column chromatograms. This PCB interference was not realized by earlier workers prior to the identification of their peaks in gas chromatograms by Jensen[4] in 1966. As has been emphasized in chapter 2, Volume I, prior to 1970, gas chromatographic methods must have overestimated DDT in the environmental samples.

Since 1970, the awareness of the environmental hazard caused by PCBs has resulted in much work — both environmental monitoring and toxicity evaluation. The reader is referred to the excellent reviews on the chemistry, properties, and environmental behavior of PCBs that have appeared in the recent past.[10,12-15] The purpose of this chapter is only to underscore the possible interference of PCBs in the analysis of DDT-group compounds and to describe the methods of separation of these two types of chemicals. As has been pointed out by Peterson and Guiney,[16] fish have attained the dubious honor of being the most frequently cited PCB-contaminant problem.

The environmental behavior and the extent of uptake of PCBs by fish are controlled by the chemistry of the PCB molecule. The lower-chlorinated PCBs are more water-soluble and easily desorbed from sediments and suspended particles.[16,17] Also, they are easily metabolized by fish and eliminated faster than those isomers with a higher chlorine content. The latter are poorly metabolized by the biota and hence retained for longer periods.

II. ANALYTICAL METHODS

Jensen arrived at the conclusion that the then unidentified peaks belonged to PCBs, after examining the mass spectra of the cleaned-up eluate of the fat from a white-tailed eagle.[18] Koeman et al.[19] also employed mass spectroscopy to identify the PCB peaks. Holden and Marsden[20] used an alumina column followed by a silica column to separate the OC pesticides from PCBs. Reynolds[21] separated PCBs from most other OC compounds (except aldrin, heptachlor, and DDE) using a Florisil® column, and hexane and hexane-ether fractions. A sililic acid-celite column was used for the separation of PCBs from DDT and its analogs.[22] Miles[23] converted DDT and its metabolites to dichlorobenzophenones, leaving the PCBs unaffected, Reinke et al.[24] used a TLC (thin layer chromatography) procedure for separating OC pesticides and PCBs. The resistance of PCBs to strong acid[25] or alkali treatment is taken advantage of in other methods of separation of PCBs from OC pesticides. Improved acid or alkaline hydrolysis of OC pesticides, followed by perchlorination with H_2O_2, was described.[26] The perchlorination technique was improved and altered to facilitate better quantification of PCBs.[27] In the recent past, capillary gas chromatography is being increasingly used to resolve many more isomers.[28-30]

III. TOXICITY OF PCBs TO FISH

The acute toxicity of PCBs to fish is low.[31] Under static conditions, their acute toxicity ranged from 1.2 to 61 mg/ℓ to various species of No. American freshwater fish.[32] The 96-h LC 50 of PCBs (Aroclor® 1254) to 22-day-old fry of deep water ciscoes was >10 mg/ℓ.[33]

The chronic toxicity of PCBs, as judged by their bioconcentration potential, is relatively high. Like many other OC compounds, PCBs are bioconcentrated to a considerable extent. A few representative bioconcentration factors, by different species of fish, are shown in Table 1. The extent of bioaccumulation of PCBs did not differ

Table 1

BIOCONCENTRATION OF PCBS (COMMERCIAL FORMULATION) BY
DIFFERENT SPECIES OF FISH

Compound	Species	Conc in water ($/l$)	Duration of exposure (days)	BCF	Ref.
PCBs (?)	Brown trout	11 ng	57	100,000 — 180,000	70
Aroclor® 1254	Brook trout fry	13 μg	118	40,000 — 47,000	5
	Juvenile turbot	580 ng	15	1,000 — 10,000	71
Aroclor® 1248	Male fathead minnow	3 μg	250	63,000	72
		2.1 μg	250	166,000	72
	Female fathead minnow	3 μg	250	120,000	72
	Female fathead minnow	2.1 μg	250	270,000	72
Aroclor® 1254	Channel catfish	5.8 μg	77	56,370	32
	Channel catfish	2.4 μg	77	61,390	32
Aroclor® 1016	Pinfish	1 μg	28	25,000	73

substantially among the species of similar lipid content and size.[34] Usually the highest concentration of PCBs was found in the liver.[35] The accumulation from water was more than that from food.[35] Long duration of exposure in water was necessary for the attainment of steady-state conditions.[16] After steady-state conditions were attained, the residues were accumulated at approximately the same rate at which the fish grew. The dependence of the uptake and metabolism of different isomers on the extent of chlorination has already been mentioned. While the effect of the degree of chlorination on the metabolism in fish has been studied, the effect of chlorine position has not been studied.[16] Peterson and Guiney[16] considered that the availability of more than two adjacent unsubstituted carbon atoms appears necessary for an appreciable rate of PCB metabolism. Although soon after their identification in the environmental samples, PCBs were considered to be nonbiodegradable, later it became clear that lower-chlorinated isomers were relatively easily eliminated.

The metabolism of chlorobiphenyls was recently reviewed.[36] Hutzinger et al.[37] reported that rat and pigeon could metabolize mono-, di-, and tetrachlorobiphenyl isomers, but rainbow trout could not. However, in vitro and in vivo formation of polar metabolites of the different isomers of PCBs and the ability of different species of fish to metabolize PCBs, albeit to a limited extent, were demonstrated.[38-41]

Unlike the DDT-group compounds, the PCBs are inducers of the mixed function oxidases (MFOs) (see chapter 3, volume I). Trout that were fed Aroclor® 1254 had significantly higher MFO activity than the controls. Concentration of microsomal protein, cytochrome p-450, and *O*-de-ethylase activity were markedly increased.[42] PCB replacement compound (XFS-4169L) also increased cytochrome p-450, but not *O*-de-ethylase activity. Aroclor® 1254 induced significantly the ethoxy coumarin-*O*-de-ethylase, but not aniline hydroxylase activity and the content of cytochrome p-450 and microsomal protein in the livers of trout.[43] In another study, Aroclor® 1254 was found to increase the hepatic ethoxyresorufin *O*-de-ethylase (EROD) activity and ethoxycoumarin-*O*-de-ethylase (ECOD) activity in trout. Benzphetamine-*N*-demethylase activity (used to distinguish cytochrome p-450 from p-448 in mammals[44] was not induced in trout by Aroclor® 1254. PCB replacement compounds were also found capable of inducing EROD activity, but not always ECOD activity.[45] In yet another study, Addison and associates[46] showed that microsomal *O*-de-ethylases in cod were induced by Aroclor® 1254 but not by Aroclor® 1016. Franklin et al.[47] showed that inducibility of MFO activity depends on the chemical structure; noncoplaner hexa- and heptabro-mobiphenyl did not induce MFO activity in rainbow trout, whereas coplanar isomers could.[47] Such elevated MFO activity was recorded in the mullet liver following the feeding of a commercial mixture of PCB;[48] aniline hydroxylase activity and cytochrome b_s activity were considered to be more sensitive indicators of such inducement than aminopyrine-N-demethylase activity and cytochrome p-450 concentration.

In view of the report that the blood PCB levels of persons increased within 3 to 5 hr after consuming PCB-contaminated fish,[49] it would be interesting to discuss the levels of PCB contamination in natural populations of fish. Various species of fish from Escambia Bay, Fla. had very high concentration of PCBs (up to 184 mg/kg) in different tissues.[50] In a base line study in 1971, Veith[51] recorded 2.7 to 15 mg/kg in Lake Michigan fish. PCB concentrations in Cayuga Lake trout increased with age and were 10 to 30 mg/kg in fish that were 7 to 12 years old.[52] In the U.S., very high concentrations of PCBs (up to 100mg/kg), were recorded in the fish of the Hudson River, salmonids of Lake Ontario, and Cayuga Lake trout.[53] In most of these studies, the PCB levels of fish were higher than the levels of the DDT group.[54,51] Subsequent to the banning of the use of persistent compounds in the Western countries, the levels of PCBs have not been declining to the same extent, as is the case with OC pesticides. Between 1972 and 1976, the PCB levels in Dutch coastal fishery products remained steady.[55] In composite fish samples collected from major watersheds of the U.S. in 1976 as part of the National Pesticide Monitoring Program, the highest level of PCBs recorded was 140 mg/kg.[56] An increase in the PCB levels in the biota of the Aegean Sea was recorded between 1975 and 1979.[57]

The sublethal effect of PCBs, especially on reproductive success and survival of the developmental stages of fish was discussed in chapter 6.

IV. OTHER COMPOUNDS

Not only PCBs, but other compounds too, may appear along with the OC pesticides on EC-GC chromatograms. Some of these compounds, like the phthalate esters, have been recorded in environmental samples,[58,59] at levels much higher than those of the OC compounds, but others like the chlorinated dibenzo dioxins were not usually encountered.[60] Chlorinated terphenyls (Aroclor® 5460),[61] chlorinated paraffin,[62] and dibenzofurans (PCDFs)[63] have been recently reported in environmental samples. Improved clean-up procedures and analytical techniques are likely to bring to light many other hitherto unsuspected pollutants that have a potential for bioaccumulation. For instance, the highly toxic dioxins (TCDD) reported to be an impurity in the herbicide 2,4,5-T,[64] have been recently reported in fresh water fish[65-67] Polybrominated biphenyl esters (PBBE) were reported from fish in Sweden.[68] As has been emphasized by Zitko,[69] any compound produced on an industrial scale ultimately reaches the aquatic environment. Zitko reviewed the different classes of industrial organic chemicals that are likely to persist, and hence are likely to be an environmental hazard.

V. CONCLUSIONS

The possible environmental hazards of PCBs were not recognized until 35 years after their large-scale industrial use began. The earlier gas chromatographic studies on pesticide residues in environmental samples would have overestimated DDT, because of the close agreement between the PCB and OC pesticide peaks in the gas chromatograms. In the last 15 years, suitable methods have been developed to separate PCBs and DDT. A simultaneous study of PCBs and the DDT group compounds is advisable because of the similarity of their chemical structure and properties and hence, environmental behavior.

REFERENCES

1. Holden, A. V., Organochlorine insecticide residues in salmonid fish, *J. Appl. Ecol.*, 3, 45, 1966.
2. Holmes, D. C., Simmons, J. H., and Tatton, J. O'G., Chlorinated hydrocarbons in British wildlife, *Nature*, 216, 227, 1967.
3. Holden, A. V., Organochlorines — an overview, *Mar. Pollut. Bull.*, 12, 110, 1981.
4. Jensen, S., Report of a new chemical hazard, *New Sci.*, 32, 612, 1966.
5. Mauck, W. L., Mehrle, P. M., and Mayer, F. L., Effects of the polychlorinated biphenyl Aroclor® 1254 on the growth, survival, and bone development in brook trout *(Salvelinus fontinalis)*, *J. Fish. Res. Board Can.*, 35, 1084, 1978.
6. Anon., The rising clamor about PCB's, *Environ. Sci. Technol.*, 10, 122, 1976.
7. Stalling, D. L. and Mayer, F. L., Jr., Toxicities of PCBs to fish and environmental residues, *Environ. Health Perspect.*, 1, 159, 1972.
8. Garvey, C., Controlling PCB's—a new approach, *EPA J.*, 25, 1981.
9. Umeda, G., PCB poisoning in Japan, *Ambio*, 1, 132, 1972.
10. Reynolds, L. M., Pesticide residue analysis in the presence of polychlorobiphenyls (PCB's), *Residue Rev.*, 34, 27, 1971.
11. Frank, R. and Braun, H. E., Residues from past uses of organochlorine insecticides and PCB in waters draining eleven agricultural watersheds in southern Ontario, Canada, 1975—1977, *Sci. Total Environ.*, 20, 255, 1981.
12. Hutzinger, O., Safe, S., and Zitko, V., *Chemistry of PCB's*, CRC Press, Boca Raton, Fla., 1974.
13. Peakall, D. B., Polychlorinated biphenyls: occurrence and biological effects, *Residue Rev.*, 44, 1, 1972.
14. Gustafson, C. G., PCB's—prevalent and persistant, *Environ. Sci. Technol.*, 4, 814, 1970.
15. Cairns, Th. and Siegmund, E. G., Regulatory history and analytical problems, *Anal. Chem.*, 53, 1183A, 1981.
16. Peterson, R. E. and Guiney, P. D., Disposition of polychlorinated biphenyls in fish, in *ACS Symposium Series, No. 99*, Khan, M. A. Q., Lech, J. J., and Menn, J. J., Eds., American Chemical Society, Washington, D. C., 1979, 21.
17. Halter, M. T. and Johnson, H. E., A model system to study the desorption and biological availability of PCB in hydrosoils, in *Aquatic Toxicology and Hazard Evaluation*, ASTM STP 634, Mayer, F. L. and Hamelink, J. L., Eds., American Society for Testing and Materials, Philadelphia, 1977, 178.
18. Jensen, S., The PCB story, *Ambio*, 1, 123, 1972.
19. Koeman, J. H., Ten Noever De Brauw, M. C., and De Vos, R. H., Chlorinated biphenyls in fish, mussels and birds from the River Rhine and the Netherlands coastal area, *Nature*, 221, 1126, 1969.
20. Holden, A. V. and Marsden, K., Single-stage clean-up of animal tissue extracts for organochlorine residue analysis, *J. Chromatogr.*, 44, 481, 1969.
21. Reynolds, L. M., Polychlorobiphenyls (PCB's) and their interference with pesticide residue analysis, *Bull. Environ. Contam. Toxicol.*, 4, 128, 1969.
22. Armour, J. A. and Burke, J. A., Method for separating polychlorinated biphenyls from DDT and its analogs, *J. Assoc. Off. Anal. Chem.*, 53, 761, 1970.
23. Miles, J. R. W., Conversion of DDT and its metabolites to dichlorobenzophenones for analysis in the presence of PCB's, *J. Assoc. Off. Anal. Chem.*, 55, 1039, 1972.
24. Reinke, J., Uthe, J. F., and O'Brodovich, H., Determination of polychlorinated biphenyls in the presence of organochlorine pesticides by thin-layer chromatography, *Environ. Lett.*, 4, 201, 1973.
25. *Pesticide Analytical Manual*, Food and Drug Administration, Washington, D. C., 1977, 1.
26. Jan, J. and Malnersic, S., Determination of PCB and PCT residues in fish by tissue acid hydrolysis and destructive clean-up of the extract, *Bull. Environ. Contam. Toxicol.*, 19, 772, 1978.
27. Stratton, C. L., Mark Allan, J., and Whitlock, S. A., Advances in the application of the perchlorination technique for the quantitation and confirmation of polychlorinated biphenyls (PCBs), *Bull. Environ. Contam. Toxicol.*, 21, 230, 1979.
28. Ballschmiter, K. and Zell, M., Analysis of polychlorinated biphenyls (PCB) by glass capillary gas chromatography, *Fresenius Z. Anal. Chem.*, 31, 302, 1980.

29. Kerkhoff, M. A. T., de Vries, A., Wegman, R. C. C., and Hofstee, A. W. M., Analysis of PCBs in sediments by glass capillary gas chromatography, *Chemosphere,* 11, 165, 1982.
30. Duinker, J. C., Hillebrand, M. T. J., Palmork, K. H., and Wilhelmsen, S., An evaluation of existing methods for quantitation of polychlorinated biphenyls in environmental samples and suggestions for an improved method based on measurement of individual components, *Bull. Environ. Contam. Toxicol.,* 25, 956, 1980.
31. Hansen, D. J., Parrish, P. R., and Forester, J., Aroclor 1016: toxicity to and uptake by estuarine animals, *Environ. Res.,* 7, 363, 1974.
32. Mayer, F. L., Mehrle, P. M., and Sanders, H. O., Residue dynamics and biological effects of polychlorinated biphenyls in aquatic organisms, *Arch. Environ. Contam. Toxicol.,* 5, 501, 1977.
33. May Passino, D. R. and Kramer, J. M., Toxicity of arsenic and PCBs to fry of deepwater ciscoes (*Coregonus*), *Bull. Environ. Contam. Toxicol.,* 24, 527, 1980.
34. Skea, J. C., Simonin, H. A., Dean, H. J., Colquhoun, J. R., Spragnoli, J. J., and Veith, G. D., Bioaccumulation of Aroclor 1016 in Hudson River fish, *Bull. Environ. Contam. Toxicol.,* 22, 332, 1979.
35. Narbonne, J. F., Accumulation of polychlorinated biphenyl (Phenoclor DP6) by estuarine fish, *Bull. Environ. Contam. Toxicol.,* 22, 60, 1979.
36. Sundström, G., Hutzinger, O., and Safe, S., The metabolism of chlorobiphenyls—a review, *Chemosphere,* 5, 267, 1976.
37. Hutzinger, O, Nash, D. M., Safe, S., DeFreitas, A. S. W., Norstrom, R. J., Wildish, D. J., and Zitko, V., Polychlorinated biphenyls: metabolic behavior of pure isomers in pigeons, rats, and brook trout, *Science,* 178, 312, 1972.
38. Sanborn, J. R., Childers, W. F., and Metcalf, R. L., Uptake of three polychlorinated biphenyls, DDT, and DDE by the green sunfish, *Lepomis cyanellus Raf., Bull. Environ. Contam. Toxicol.,* 13, 209, 1975.
39. Melancon, M. J., Jr. and Lech, J. J., Isolation and identification of a polar metabolite of tetrachlorobiphenyl from bile of rainbow trout exposed to ^{14}C-tetrachlorobiphenyl, *Bull. Environ. Contam. Toxicol.,* 15, 181, 1976.
40. Hinz, R. and Matsumura F., Comparative metabolism of PCB isomers by three species of fish and the rat, *Bull. Environ. Contam. Toxicol.,* 18, 631, 1977.
41. Herbst, E., Scheunert, I., Klein, W., and Korte, F., Uptake and conversion of 2,5,4'-trichlorobiphenyl-^{14}C, 2,4,6,2,4'-pentachlorobiphenyl-^{14}C and chloroalkylene-9-^{14}C by goldfish after a single water treatment, *Chemosphere,* 7, 221, 1978.
42. Addison, R. F., Zinck, M. E., Willis, D. E., and Darrow, D. C., Induction of hepatic mixed function oxidases in trout by polychlorinated biphenyls and butylated monochlorodiphenyl ethers, *Toxicol. Appl. Pharmacol.,* 49, 245, 1979.
43. Addison, R. F., Zinck, M. E., and Willis, D. E., Induction of hepatic mixed-function oxidase (MFO) enzymes in trout (*Salvelinus fontinalis*) by feeding Aroclor® 1254 or 3-methylcholanthrene, *Comp. Biochem. Physiol.,* 61, 323, 1978.
44. Lech, J. J., Vodicnik, M. J., and Elcombe, C. R., Induction of monooxygenase activity in fish, in *Aquatic Toxicology,* Weber, L. J., Ed., Raven Press, New York, 1982, 107.
45. Addison, R. F., Zinck, M. E., Willis, D. E., and Wrench, J. J., Induction of hepatic mixed function oxidase activity in trout (*Salvelinus fontinalis*) by Aroclor 1254 and some aromatic hydrocarbon PCB replacements, *Toxicol. Appl. Pharmacol.,* 63, 166, 1982.
46. Hansen, P. -D., Addison, R. F., and Willis, D. E., Hepatic microsomal o-de-ethylases in cod (*Gadus morhua*): their induction by Aroclor 1254 but not by Aroclor 1016, *Comp. Biochem. Physiol.,* 74, 173, 1983.
47. Franklin, R. B., Vodicnik, M. J., Elcombe, C. R., and Lech, J. J., Alterations in hepatic mixed-function oxidase activity of rainbow trout after acute treatment with polybrominated biphenyl isomers and firemaster BP-6, *J. Toxicol. Environ. Health,* 7, 817, 1981.
48. Narbonne, J. F. and Gallis, J. L., *In vivo* and *in vitro* effect of Phenochlor DP6 on drug metabolizing activity in mullet liver, *Bull. Environ. Contam. Toxicol.,* 23, 338, 1979.
49. Kuwabara, K., Yakushiji, T., Watanabe, I., Yoshida, S., Yoyaka, K., and Kunita, N., Increase in the human blood PCB levels promptly following ingestion of fish containing PCBs, *Bull. Environ. Contam. Toxicol.,* 21, 273, 1979.
50. Duke, T. W., Lowe, J. I., and Wilson, A. J., Jr., A polychlorinated biphenyl (Aroclor 1254) in the water, sediment, and biota of Escambia Bay, Florida, *Bull. Environ. Contam. Toxicol.,* 5, 171, 1970.
51. Veith, G. D., Baseline concentrations of polychlorinated biphenyls and DDT in Lake Michigan fish, 1971, *Pestic. Monit. J.,* 9, 21, 1975.
52. Bache, C. A., Serum, J. W., Youngs, W. D., and Lisk, D. J., Polychlorinated biphenyl residues: accumulation in Cayuga Lake trout with age, *Science,* 177, 1191, 1972.

53. Spagnoli, J. J. and Skinner, L. C., PCB's in fish from selected waters of New York State, *Pestic. Monit. J.*, 11, 69, 1977.

54. Claeys, R. R., Caldwell, R. S., Cutshall, N. H., and Holton, R., Chlorinated pesticides and polychlorinated biphenyls in marine species, Oregon/Washington coast, 1972, *Pestic. Monit. J.*, 9, 2, 1975.

55. Hagel, P. and Tuinstra, L. G. M. Th., Trends in PCB contamination in Dutch coastal and inland fishery products 1972—1976. *Bull. Environ. Contam. Toxicol.*, 19, 671, 1978.

56. Veith, G. D., Kuehl, D. W., Leonard, E. N., Puglisi, F. A., and Lemke, A. E., Polychlorinated biphenyls and other organic chemical residues in fish from major watersheds of the United States, 1976, *Pestic. Monit. J.*, 13, 1, 1979.

57. Kilikidis, S. D., Psomas, J. E., Kamarianos, A. P., and Panetsos, A. G., Monitoring of DDT, PCBs, and other organochlorine compounds in marine organisms from the North Aegean Sea, *Bull. Environ. Contam. Toxicol.*, 26, 496, 1981.

58. Giam, C. S. and Atlas, E., Accumulation of phthalate ester plasticizers in Lake Constance sediments, *Naturwissenschaften*, 67, 508, 1980.

59. Giam, C. S., Atlas, E., Chan, H., and Neff, G., Estimation of fluxes of organic pollutants to the marine environment phthalate plasticizer concentration and fluxes, *Rev. Int. Oceanogr. Med.*, 47, 1977.

60. Zitko, V., Absence of chlorinated dibenzodioxines and dibenzofurans from aquatic animals, *Bull. Environ. Contam. Toxicol.*, 7, 105, 1972.

61. Addison, R. F., Fletcher, G. L., Ray, S., and Doane, J., Analysis of a chlorinated terphenyl (Aroclor 5460) and its deposition in tissues of cod (*Gadus morhua*), *Bull. Environ. Contam. Toxicol.*, 8, 52, 1972.

62. Svanberg, O., Bengtsson, B. -E., and Linden, E., Paraffins—a case of accumulation and toxicity to fish, *Ambio*, 7, 65, 1978.

63. Lake, J. L., Rogerson, P. F., and Norwood, C. B., A polychlorinated dibenzofuran and related compounds in an estuarine ecosystem, *Environ. Sci. Technol.*, 15, 549, 1981.

64. Ahling, B., Lindskog, A., Jansson, B., and Sundström, G., Formation of polychlorinated dibenzo-p-dioxins and dibenzofurans during combustion of a 2,4,5,-T formulation, *Chemosphere*, 6, 461, 1977.

65. Harless, R. L., Oswald, E. O., Lewis, R. G., Dupuy, A. E., Jr., McDaniel, D. D., and Tai, H., Determination of 2,3,7,8-tetrachlorodibenzo-p-dioxin in freshwater fish, *Chemosphere*, 11, 193, 1982.

66. Yamagishi, T., Miyazaki, T., Akiyama, K., Morita, M., Nakagawa, J., Horii, S., and Kaneko, S., Polychlorinated dibenzo-p-dioxins and dibenzofurans in commercial diphenyl ether herbicides, and in freshwater fish collected from the application area, *Chemosphere*, 10, 1137, 1981.

67. O'Keefe, P., Meyer, C., Hilker, D., Aldous, K., Jelus-Tyror, B., Dillon, K., Donnelly, R., Horn, E., and Sloan, R., Analysis of 2,3,7,8-tetrachlorodibenzo-p-dioxin in Great Lakes fish, *Chemosphere*, 12, 325, 1983.

68. Andersson, O. and Blomkvist, G., Polybrominated aromatic pollutants found in fish in Sweden, *Chemosphere*, 10, 1051, 1981.

69. Zitko, V., Potentially persistent industrial organic chemicals other than PCB, in *Ecological Toxicology Research*, McIntyre, A. D. and Mills, C. F., Eds., Plenum Press, New York, 1975.

70. Spigarelli, S. A., Thommes, M. M., and Prepejchal, W., Thermal and metabolic factors affecting PCB uptake by adult brown trout, *Environ. Sci. Technol.*, 17, 88, 1983.

71. Courtney, W. A. M. and Langston, W. J., Accumulation of polychlorinated biphenyls in turbot (*Scophthalmus maximus*) from sea water sediments, *Helgol. Meeresunters.*, 33, 333, 1980.

72. DeFoe, D. L., Veith, G. D., and Carlson, R. W., Effects of Aroclor® 1248 and 1260 on the fathead minnow (*Pimephales promelas*), *J. Fish. Res. Board Can.*, 35, 997, 1978.

73. Bloom, H., Taylor, W., Bloom, W. R., and Ayling, G. M., Organochlorine pesticide residues in animals of Tasmania, Australia — 1975—77, *Pestic. Monit. J.*, 13, 99, 1979.

Chapter 8

ENVIRONMENTAL HAZARD EVALUATION AND PREDICTION

I. INTRODUCTION

Aquatic toxicity testing is perhaps 40 to 50 years old, but "Aquatic Toxicology" as a separate branch took roots only after interest in environmental problems was aroused, following the publication of Rachel Carson's *Silent Spring*.[1] From the simple tests conducted to study the toxic effects of chemicals on aquatic organisms, the principal aim of aquatic toxicologists has now shifted to the evaluation of the hazard that may be caused by the continued use of a substance, for it is now well established that all man-made chemicals, when produced on an industrial scale, eventually reach the aquatic environment. The ultimate goal is the protection of the diverse aquatic organisms and whole communities from the dire effects of man-made chemicals.[2] Aquatic toxicity testing is a means to achieve this end, and differs from mammalian toxicity testing, where the effects of chemicals on rats, mice, and guinea pigs are studied with a view to extrapolating such studies to human beings. Although it has been suggested that aquatic systems may be used as tools in human health evaluation,[3] Shigan[4] concluded that the use of the toxic response of lower animals to predict the toxicity of substances to higher animals is neither possible nor profitable. Hence, in aquatic toxicology, fish are studied to protect only fish.[2,5]

The environmental hazard of any chemical is the result of its accumulation in the environment at specific locations and in various media during manufacture, application, or dissipation.[6] In such an event, to lessen the effect of pesticides on aquatic organisms, it would be ideal if the entry of pesticides into the aquatic environment could be prevented; however, this is not possible. The conflicting interest between the need to boost agricultural and timber production and to protect the produce from pest attack, and the desire to maintain a clean environment, is one of the tragedies of the 20th century.

Aquatic toxicity testing is a means of identifying the risk to the nontarget organisms, arising from the continued use of pesticides and other chemicals. The testing of the toxicity of a vast array of pesticides to the diverse organisms that inhabit the aquatic world, however, is an impossible task, if every species has to be protected from the effect of every pesticide that may eventually find its way into the aquatic environment. Besides, aquatic organisms can tolerate occasional adverse effects to a certain extent, and hence the protection of all species, at all times, is not necessary.[7] It would suffice if a large number of appropriate taxa (from a variety of taxonomic and functional groups) are protected, unless a commercially, recreationally, or socially important species is very sensitive.[7] It is in this direction that hazard evaluation and predictive toxicology play a useful and important role.

II. HAZARD EVALUATION AND PREDICTION

Since the risk posed to the majority of aquatic organisms seems to be controlled by the truly dissolved concentration of pesticides, hazard evaluation depends on identifying the sources and sinks of these chemicals in the aquatic environment.[8] Predicting when, where, and to what extent fish in particular, and other organisms in general, accumulate chemicals, has ecological and economic significance.[9] Consequently, an important question arises as to whether it would be possible to forecast the environmental safety or otherwise of a pesticide, even before its extensive use.

The model ecosystem, devised by Metcalf et al.,[10] and discussed in Chapter 3, Volume I, has proved very useful in evaluating the environmental behavior of pesticides and other chemicals. Besides, in the recent past it has been possible to predict theoretically the environmental fate of chemicals. The characteristics of a substance for which testing is required to evaluate its environmental fate include persistence, mobility in the different compartments of the environment, acute toxicity, sublethal toxicity, and chronic toxicity (reproductive impairment). According to Stern and Walker,[11] studies on water solubility, octanol-water partition coefficient, volatility, absorption by natural solids, and desorption and leaching of a chemical help identify the different compartments of the environment in which a chemical may accumulate. With the additional knowledge of the precise amounts of a pesticide produced and used, and the extent of its chemical-, photo-, and biodegradability, it would be possible to predict the actual amounts that may eventually reach and remain in the aquatic environment. Recently, Wauchope[12] discussed the protocols necessary for testing the potential of pesticides that may be washed into streams and other aquatic bodies, following their application to agricultural fields and forests. Neely and Mackay[13] described a modeling process to access the distribution of chemicals in the environment. Using the vapor pressure, water solubility, and molecular weight of a chemical, and combining them with its rate of degradation, Neely[14] devised a model to estimate the relative persistence of a chemical. On the basis of this model it would be possible to identify the chemicals that may be considered as priority pollutants. Frische et al.[15] and Klopffer et al.[16] also discussed the various physicochemical properties of chemicals that would be useful in predicting their environmental behavior. Thus, it is now possible to identify pesticides and other chemicals that can be potentially hazardous to the aquatic life, without resorting to extensive experimental studies.

With a view to avoiding undue risk to the environment and human health, measures like the Toxic Substances Control Act (TSCA) in the U.S., the Chemical Substances Control Law in Japan, and similar laws in other countries have been promulgated.[17,18] Since it is humanly impossible to test every new chemical that is being added to the vast number of already existing ones, many of these legislative measures envisage a system of tiered testing. One such system, ranging from base set of level 0 (involving degradability and acute toxicity) to level 2 (involving prolonged tests), being followed by the member states of the European Community, is shown in Table 1.[19] At different levels, the risks involved and the potential hazard of a chemical, including that of all new pesticides, is evaluated and if the risks are low or negligible, further testing is discontinued and the chemical may be marketed. On the other hand, if a potential risk is indicated, further extensive tests of a complicated nature are conducted and the advisability of permitting or preventing the manufacture and marketing of such a pesticide can be made. For instance, if a pesticide is readily degradable, has low toxicity, or is produced only in small quantities, further testing would be wasteful. If on the other hand the compound is to be widely used, or a potential for bioaccumulation is indicated, further extensive testing is required and perhaps a decision to disallow its environmental use has to be taken. Similarly, as illustrated in Figure 1, if the expected environment concentrations are well below the highest concentration that produced no adverse effect, the point at which the sequential testing may be discontinued can readily be identified. As the environmental concentration approaches the no adverse biological effects threshold, not only greater numbers of tests would be required to identify the degree of overlap between the two, but should the overlap be considerable, it would be necessary to restrict or ban the use of that pesticide.[20] Branson et al.[21] discussed a similar tiered system of testing for effluent monitoring.

Theoretical approaches to predicting the likely hazard of a pesticide (or chemical) are comparatively new and less than a decade old. Branson[22] cautioned against an

Table 1

SEQUENTIAL SCHEME FOR ECOTOXICITY TESTING

Level 0 (basic dossier) — mandatory
  Degradability
    Abiotic degradability
    Biodegradability
  Acute toxicity (lethal effects)
    LC 50 with *Daphnia*
    LC 50 with fish
Level 1 (if quantities put on the market = 10 tons/year or a total amount of 50 tons: desirable; if quantities
  put on the market = 100 tons/year or a total amount of 500 tons: mandatory)
  Growth inhibition test in algae
    Species to be chosen
  Prolonged toxicity study with *Daphnia magna*
    Duration: 21 days
    With determination of effects on reproduction and of lethal effects
  Prolonged toxicity study with fish (e.g., *Oyrzias, Jordanella,* etc.)
    At least a period of 14 days
    Species to be chosen
    With determination of the "threshold level"
  Test for species accumulation
    One species, preferably fish (e.g., *Poecilia reticulata)*
  Prolonged degradability study
    Dynamic test with lower conc and with another inoculum than used in Step 0
  Test with superior plant
  Test with earthworms
Level 2 (to be carried out if the quantity placed on the market reaches 1000 tons/year or a total of 5000 tons,
  and if there are no strong reasons to the contrary) tests must cover:
  Additional testing for accumulation, degradation, and mobility
  Prolonged toxicity testing with fish (including reproduction)
  Additional toxicity study (acute and subacute) with birds (quail) if accumulation factor is >100
  Additional toxicity studies with other organisms (need should be indicated and choice of organisms on
    chemicals used)
  Study on adsorption-desorption, if substance is not particularly degradable

From Smeets, J., *Ecotoxicol. Environ. Saf.,* 4, 103, 1980. With permission.

unscientific way of designating certain chemicals as priority pollutants, and suggested that priority should be accorded to those with the least margin of safety, i.e., those chemicals for which there is not much difference between the no observable effect concentration (NOEC) and the ambient exposure concentration (the area where the dotted lines overlap in Figure 1). While the NOEC can be calculated following embryo-larval[23] or chronic life cycle tests,[24,25] the expected ambient concentrations could be estimated from a nomogram that requires information on any two of three factors, viz., the environmental release rate, ratio of dissipation to bioconcentration potential, and ambient residues in fish.

Maki[26] compared 13 different programs of hazard evaluation schemes that include some type of tiered system of testing. Most of these systems agree with one another, with respect to test requirements in the early tiers. A high degree of variation was observed at higher levels. Maki suggested that basing decisions solely on the volume of use or extent of distribution, regardless of its inherent toxicity, is often misleading. He suggested that a comparison of LC 50 with the expected environmental concentration would be an objective criterion for hazard assessment.

Cairns[20] also emphasized the limitations of such a prediction, despite its certain advantage in estimating the hazard of a chemical before it is manufactured and marketed. First, an extrapolation has to be made from a few species to many, on the basis of the study of a limited number of environmental conditions. Second in the absence of feed-

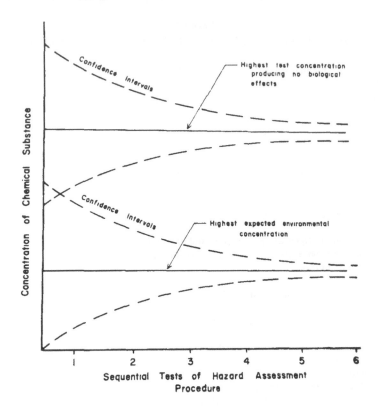

FIGURE 1.   Diagrammatic representation of a sequential hazard-assessment procedure demonstrating increasingly narrow confidence limits for estimates of no biological effect concentration and actual expected environmental concentration. (From *Estimating the Hazard of Chemical Substances to Aquatic Life,* ASTM STP 657, Cairns, J., Jr., Dickson, K. L., and Maki, A. W., Eds., American Society for Testing and Materials, Philadelphia, 1978. With permission.)

back from the environment, the validity of the predictions cannot be accurately gauged.

Quantitative Structure Activity Relationship (QSAR) studies have proved to be another promising method of predicting the fate and behavior of a chemical in the environment. The usefulness of QSAR studies in evaluating new chemicals, and the limitations of such correlations have been discussed in detail in Chapter 3, Volume I. In general, the chemical structure and the physicochemical properties deriving from the chemical structure, have proved reliable in predicting environmental behavior and extent of uptake of xenobiotic chemicals by the aquatic organisms. Derivation of the Application Factors (AFs) following the calculation of the Maximum Acceptable Toxicant Concentration (MATC) has also been a useful means of identifying the levels at which a species may be affected in the environment. Recently, Buikema et al.[27] have reviewed the usefulness of safe concentrations in predicting the levels of a toxicant that are unlikely to harm a population or ecosystem. The methods of calculation of MATC and AFs were reviewed in Chapter 1.

In the U.S., water quality criteria for several compounds have been developed with a view to protecting the aquatic life. These criteria are primarily based on data on the acute toxicity to aquatic organisms and the maximum concentration which will not cause any adverse effect during long-term exposure. Generally, concentrations of toxicants that do not cause unacceptable adverse effects on animals, also do not cause adverse effects on aquatic vegetation.[7] Hence, the criteria are based on the data on the toxicity of a compound to animals. Guidelines for deriving numerical National Water Quality Criteria for the protection of aquatic life in the U.S. were revised in 1983 and the method of deriving these criteria was described.[7]

III. BIOLOGICAL MONITORING

Following Hellawell,[28] Cairns[20] has defined the three terms — survey, surveillance, and monitoring as follows:

- Survey: an exercise in which a set of standardized observations (or replicate samples) is taken from a station (or stations) within a short period of time to furnish qualitative or quantitative descriptive data
- Surveillance: a continued program of surveys systematically undertaken to provide a series of observations in time
- Monitoring: surveillance undertaken to ensure that previously formulated standards are being met

Together, these constitute "Biological Monitoring", which is now recognized as an important means of identifying adverse environmental changes (also referred to as "Environmental Perturbation" by some authors).

Changed brain acetylcholinesterase (AChE) levels of fish in stressed environments have been used as a tool for biological monitoring.[29-33] The inhibition of fish brain AChE following exposure to organophosphate (OP) compounds was discussed in Chapter 6. Similarly, the induction of mixed function oxidase (MFO) activity by certain classes of xenobiotic chemicals, is also a useful tool for biological monitoring (see Chapter 6). Likewise, the changes in fish cough response and respiratory movements[34,35] were considered useful in monitoring changes in the quality of the effluent receiving systems. Cairns and Gruber[36] described a method of interfacing a system that monitors the breathing rates of fish with a minicomputer, to serve as an early warning system and for detecting changes in the environmental quality. On the other hand, Sloof[37] considered the fish respiratory activity to be of limited use for detecting the potentially hazardous substances.

The foregoing discussion clearly indicates that the simple toxicity tests of earlier times have given way to Predictive Toxicology. Instead of taking corrective measures after the deterioration of the environment, the emphasis now is on preventing the entry of possibly hazardous chemicals into the environment, even by denying registration, thereby preventing the marketing of a chemical.

IV. GENERAL CONCLUSIONS

In all the previous chapters, an attempt has been made to identify the levels of environmental pollution by pesticides, and their effect on fish. The various means by which the aquatic organisms are exposed to the toxic action of pesticides, and the levels of accumulation of pesticide residues by fish are listed. The toxicity of pesticides to fish (both acute and chronic) and the various factors that modify the toxicity are discussed. Joint action of pesticides on fish is poorly understood and the theoretical premises to study this problem are not well defined. Much information has been gathered on the different types of changes in fish induced by sublethal concentrations of pesticides; however, the ecological significance of the observed changes is little understood.

Since the ultimate aim of all toxicity testing is to assess how deleterious a compound will be to the biota, present-day toxicity test protocols lay stress on preventing a disaster, rather than taking corrective action after its occurrence, as was the case with DDT and other organochlorine (OC) compounds in the late 1960s and early 1970s. Present-day aquatic toxicity testing is concerned more with the protection of diverse aquatic organisms, rather than with the enumeration of the lethal and sublethal effects of one toxicant on a single species. The immediate future of Aquatic Toxicology (including the toxicity of pesticides) lies in the realm of Predictive Toxicology, and in determining the confidence with which laboratory experiments can be extrapolated to field conditions.

REFERENCES

1. Carson, R., *Silent Spring,* Houghton Mifflin, Boston, 1962, 360.
2. Dagani, R., Aquatic toxicology matures, gains importance, *Chem. Eng. News,* 58, 18, 1980.
3. Macek, K. J., Aquatic toxicology: fact or fiction?, *Environ. Health Perspect.,* 34, 159, 1980.
4. Shigan, S. S., Methods for predicting chronic toxicity parameters of substances in the area of water hygiene, *Environ. Health Perspect.,* 13, 83, 1976.
5. Marking, L. L. and Kimerle, R. A., Summary, in *Aquatic Toxicology,* American Society for Testing and Materials, Philadelphia, 1977, 383.
6. Blau, G. E. and Neely, W. B., What constitutes an adequate model for predicting the behavior of pesticides in the environment?, *Residue Rev.,* 85, 294, 1983.
7. Stephan, C. E., Mount, D. I., Hansen, D. J., Gentile, J. H., Chapman, G. A., and Brungs, W. A., Guidelines for deriving numerical National Water Quality Criteria for the protection of aquatic organisms and their uses, PB 85-227049, National Technical Information Service, Springfield, Va., 1985.
8. Hamelink, J., Bioavailability of chemicals in aquatic environments, in *Biotransformation and Fate of Chemicals in the Environment,* Maki, A. W., Dickson, K. L., and Cairns, J. Jr., Eds., American Society of Microbiology, Washington, D.C., 1980, 56.
9. Hamelink, J. L., Fish and chemicals: the process of accumulation, *Ann. Rev. Pharmacol. Toxicol.,* 17, 167, 1977.
10. Metcalf, R. L., Sangha, G. K., and Kapoor, I. P., Model ecosystem for the evaluation of pesticide biodegradability and ecological magnification, *Environ. Sci. Technol.,* 5, 709, 1971.
11. Stern, A. M. and Walker, C. R., Hazard assessment of toxic substances: environmental fate testing of organic chemicals and ecological testing, in *Estimating the Hazard of Chemical Substances to Aquatic Life,* ASTM STP 657, Cairns, J., Jr., Dickson, K. L., and Maki, A., Eds., American Society for Testing and Materials, Philadelphia, 1978, 81.
12. Wauchope, R. D., Runoff studies and pesticide registration, in Test Protocols for Environmental Fate and Movement of Toxicants, Proc. Symp. Assoc. Off. Anal. Chemicals, Washington, D.C., 1980.
13. Neely, W. B. and Mackay, D., Evaluative model for estimating environmental fate, in *Modeling the Fate of Chemicals in the Aquatic Environment,* Dickson, K. L., Maki, A. W., and Cairns, J., Jr., Eds., Ann Arbor Science, Ann Arbor, Mich., 1982.
14. Neely, W. B., Organizing data for environmental studies, *Environ. Toxicol. Chem.,* 1, 259, 1982.
15. Frische, R., Esser, G., Schönborn, W., and Klopffer, W., Criteria for assessing the environmental behavior of chemicals: selection and preliminary quantification, *Ecotoxicol. Environ. Saf.,* 6, 283, 1982.
16. Klopffer, W., Rippen, G., and Frische, R., Physicochemical properties as useful tools for predicting the environmental fate of organic chemicals, *Ecotoxicol. Environ. Saf.,* 6, 294, 1982.
17. Draggan, S., TSCA: the US attempt to control toxic chemicals in the environment, *Ambio,* 7, 260, 1978.
18. Fujiwara, K., Japanese law on new chemicals and the methods to test the biodegradability and bioaccumulation of chemical substances, in *Analyzing the Hazard Evaluation Process,* Dickson, K. L., Maki, A. W., and Cairns, J., Jr., Eds., Water Quality Section, American Fisheries Society, Washington, D.C., 1979, 50.
19. Smeets, J., Tests and notification of chemical substances in current legislation, *Ecotoxicol. Environ. Saf.,* 4, 103, 1980.
20. Cairns, J., Jr., Estimating hazard, *BioScience,* 30, 101, 1980.
21. Branson, D. R., Armentrout, D. N., Parker, W. M., Hall, C. V., and Bone, L. I., Effluent monitoring step by step, *Environ. Sci. Technol.,* 15, 513, 1981.

22. Branson, D. R., Prioritization of chemicals according to the degree of hazard in the aquatic environment, *Environ. Health Perspect.*, 34, 133, 1980.

23. McKim, J. M., Evaluation of tests with early life stages of fish for predicing long-term toxicity, *J. Fish. Res. Board Can.*, 34, 1148, 1977.

24. Eaton, J. G., Recent developments in the use of laboratory bioassays to determine safe levels of toxicants for fish, in *Bioassay Techniques and Environmental Chemistry*, Ann Arbor Science, Ann Arbor, Mich., 1973, 107.

25. Reinert, R. D., Stone, L. J., and Willford, W. A., Effect of temperature on accumulation of methylmercuric chloride and p,p′-DDT by rainbow trout *(Salmo gairdneri)*, *J. Fish. Res. Board Can.*, 31, 1649, 1974.

26. Maki, A. W., An analysis of decision criteria in environmental hazard evaluation programs, in *Analyzing the Hazard Evaluation Process*, Dickson, K. L., Maki, A. W., and Cairns, J., Jr., Eds., Water Quality Section, American Fisheries Society, Washington, D.C., 1979, 83.

27. Buikema, A. L., Niederlehner, B. R., and Cairns, J., Jr., Biological monitoring. IV. Toxicity testing, *Water Res.*, 16, 239, 1982.

28. Hellawell, J. M., *Biological Surveillance of Rivers*, Water Research Centre, Stevenage, England, 1978.

29. Weiss, C. M., Physiological effect of organic phosphorus insecticides on several species of fish, *Trans. Am. Fish. Soc.*, 90, 143, 1961.

30. Weiss, C. M., Use of fish to detect organic insecticides in water, *J. Water Pollut. Control Fed.*, 37, 647, 1965.

31. Weiss, C. M. and Gakstatter, J. H., Detection of pesticides in water by biochemical assay, *J. Water Pollut. Control Fed.*, 36, 240, 1964.

32. Coppage, D. L., Organophosphate pesticides: specific level of brain AChE inhibition related to death in sheepshead minnows, *Trans. Am. Fish. Soc.*, 101, 534, 1972.

33. Coppage, D. L., Matthews, E., Cook, G. H., and Knight, J., Brain acetylcholinesterase inhibition in fish as a diagnosis of environmental poisoning by malathion, *O,O*-dimethyl s-(1,2-dicarbethoxyethyl) phosphorodithioate, *Pestic. Biochem. Physiol.*, 5, 536, 1975.

34. Maki, A. W., Respiratory activity of fish as a predictor of chronic fish toxicity values for surfactants, in *Aquatic Toxicology*, ASTM ATP 667, Marking, L. L. and Kimerle, R. A., Eds., American Society for Testing and Materials, Philadelphia, 1979, 77.

35. Spoor, W. A., Neiheisel, T. W., and Drummond, R. A., An electrode chamber for recording respiratory and other movements of free-swimming animals, *Trans. Am. Fish. Soc.*, 100, 22, 1971.

36. Cairns, J., Jr., and Gruber, D., Coupling mini- and microcomputers to biological early warning systems, *Bioscience*, 29, 665, 1979.

37. Sloof, W., Detection limits of a biological monitoring system based on fish respiration, *Bull. Environ. Contam. Toxicol.*, 23, 517, 1979.

Appendix I

INDEX OF COMPOUNDS MENTIONED IN THE TEXT AND THEIR CHEMICAL NAMES

Abate®	See Temephos
ABS	Alkylbenzene sulfonate
Acephate	*O,S*-Dimethyl *N*-acetylphosphoramidothioate
Akton®	*O,O*-Diethyl *O*-(2-chloro-1-[2,5-dichlorophenyl])vinyl-phosphorothioate
Aldrin	1,2,3,4,10,10-Hexachloro-1,4,4a,5,8,8a-hexahydro-1,4-endo-exo-5,8-dimethanonaphthalene
Allethrin	dl-2-Allyl-4-hydroxy-3-methyl-2-cyclopenten-1-one ester of dl *cis/trans*- 2,2-dimethyl-3-(2-methylpropenyl)-cyclopropanecarboxylic acid
Aminocarb	4-(Dimethylamino)-*m*-tolyl methylcarbamate
Anilazine	4,6-Dichloro-*N*-(2-chlorophenyl)-1,3,5-triazin-2-amine
Aquathol® K 40	Formulation of endothall; see endothall
Aroclors®	Commercial formulations of PCBs and terphenyls of differing chlorine content, marketed by Monsanto Co., U.S.
Asulam®	Methyl(4-aminobenzenesulfonyl)carbamate
Atrazine	2-Chloro-4-ethylamino-6-isopropylamino-1,3,5-triazine
Azinophos ethyl	*O,O*-Diethyl *S*-([4-oxo-1,2,3-benzotriazin-3(4H)-yl]methyl) phosphorodithioate
Azinophos methyl	*O,O*-Dimethyl *S*-([4-oxo-1,2,3-benzotriazin-3-(4H)-yl]methyl) phosphorodithioate
Azodrin®	Dimethyl phosphate of 3-hydroxy-*N*-methyl-*cis*-crotonamide
Bayer® 73	See Bayluscide®
Baygon®	See propoxur
Bayluscide®	2′,5-Dichloro-4′-nitrosalicylamide, 2-aminoethanol salt
Bensulide	*O,O*-Diisopropyl phosphorodithioate *S*-ester of *N*-(2-mercaptoethyl) benzenesulfonamide
Benzo(a)pyrene	
Bioallethrin®	See allethrin
Bioethanomethrin	
Bis(tributyltin) oxide	
Bolero®	See thiobencarb
Bromoxynil	3,5-Dibromo-4-hydroxybenzonitrile
Bromobiphenyls, di, tri, tetra	See polybrominated biphenyls
Butoxy ethyl ester of 2,4-D	See 2,4-D
Captan	*N*-(Trichloromethylthio)-4-cyclohexene-1,2-dicarboximide
Carbaryl	1-Napthyl *N*-methyl carbamate
Carbofuran	2,3-Dihydro-2,2-dimethyl-7-benzofuranyl methyl carbamate
Carbophenothion	*S*-(*p*-Chlorophenyl methylthio)*O,O*-diethyl phosphorodithioate
Chlordane, *cis*- and *trans*-	1,2,4,5,6,7,8,8-Octachloro-2,3,3a,4,7,7a-hexahydro-4,7-methanoindene

Chlordecone	Decachlorooctahydro-1,3,4-metheno-2H-cyclobuta (cd)-pentalen-2-one
Chlorobiphenylols	Hydroxylated polychlorinated biphenyls
2-Chlorophenol	
Chlorpyrifos	*O,O*-Diethyl *O*-(3,5,6-trichloro-2-pyridyl) phosphorothioate
Chlorinated dibenzodioxins	
Chlorinated dioxins	
Chlorinated terphenyls	See polychlorinated terphenyls
Clophen	See polychlorinated biphenyls
CNP	1,3,5-Trichloro-2-(4-nitrophenoxy) benzene
Coumaphos	*O,O*-Diethyl *O*-(3-chloro-4-methyl-2-oxy [2H]-1-benzopyran-7-y *l*)phosphorothioate
CPDPP	Cumylphenyl diphenyl phosphate
Cypermethrin®	(±) *α*-Cyano-3-phenoxybenzyl(±)*cis-,trans*-3-(2,2-dichlorovinyl)-2,2-dimethyl cyclopropane carboxylate
2,4-D	2,4-Dichlorophenoxyacetic acid
2,4-D Esters	Alkonalamine, butoxy ethanol, dimethyl amine, dodecyl amine, isopropyl butyl, polyethylene glycol butyl ether, tetradodecyl amine esters of 2,4-D
p,p'-DBH	bis-*p*-Chlorophenyl methanol
DBNP	Butyl ester of 2,4-D
p,p'-DBP	4-4'-Dichlorobenzophenone
DCPA	Dimethyl tetrachlorotetraphthalate
p,p'-DDA	2,2-bis-(*p*-Chlorophenyl)-acetic acid
DDD	2,2-bis(*p*-Chlorophenyl)-1,1-dichloroethane
DDE	2,2-bis(*p*-Chlorophenyl)-1,1-dichloroethylene
p,p'-DDMS	2,2-bis(*p*-Chlorophenyl)-1-chloroethane
p,p'-DDMU	2,2-bis(*p*-Chlorophenyl)-1-chloroethylene
p,p'-DDNU	1,1-bis(*p*-Chlorophenyl)ethylene
p,p'-DDOH	2,2-bis(*p*-Chlorophenyl)ethanol
p,p'-Cl DDT	1,1,1-Trichloro-2-chloro-2,2,-bis(*P*-chlorophenyl)ethane
p,p'-DDT	2,2-bis (*p*-Chlorophenyl)-1,1,1-trichloroethane
Σ DDT	Total DDT, i.e., *p,p'*-DDT+*p,p'*-DDE+*p,p'*-DDD
DDVP	See dichlorvos
Decachlorodiphenyl	See polychlorinated biphenyls
DEF®	*S,S,S*-Tributyl phorophorotrithioate
DEHP	di-2-Ethylhexyl phthalate
Delnav®	See dioxathion
Demeton	*O,O*-Diethyl *O*- [2-(ethylthio)ethyl] phosphorothioate and *O,O*-diethyl *S*- [2-(ethylthio) ethyl] phosphorothioate
Demeton-*S*-methyl	*S*-2-Ethylthioethyl *O,O*-dimethyl phosphorothioate
Dianisyl neopentane	1,1-bis(*p*-Methoxyphenyl)-2,2-dimethyl propane
Diazinon	*O,O*-Diethyl *O*- (2-isopropyl-6-methyl-4-pyrimidinyl) phosphorothioate
Dibenzofurans	
Dicamba®	2-Methoxy-3,6-dichlorobenzoic acid
Dicapthon	*O,O*-Dimethyl *O*- 2-chloro-4-nitrophenyl phosphorothioate

Dichlobenil	2,6-Dichlorobenzonitrile
Dichlofenthion	*O,O*-Diethyl *O*-(2,4-dichlorophenyl)phosphorothioate
Dichlone	2,3-Dichloro-1,4-naphthoquinone
Dichlorobenzophenone	See DBP
Dichlorvos	2,2-Dichlorovinyl dimethyl phosphate
Dicofol	1,1-bis(4-Chlorophenyl)-2,2,2-trichlorethanol
Dieldrin	1,2,3,4,10,10-Hexachloro-exo-6,7-epoxy-1,4,4a,5,6,7,8,8a-octahydro-1,4-endo-exo-5,8-dimethanonaphthalene
Diflubenzuron	*N*-{[(4-Chlorophenyl)amino]carbonyl}-2,6-difluorobenzamide
Dimetilan	1-Dimethylcarboamoyl-5-methylpyrazol-3-yl-dimethylcarbamate
Dimethoate	*O,O*-Dimethyl *S*-(*N*-methylcarbamoylmethyl) phosphorodithioate
Dimethrin	2,4-Dimethylbenzyl-2,2-dimethyl-3-(2-methylpropenyl)cyclopropanecarboxylate
Dimilin®	See diflubenzuron
Dinitramine	N^3,N^3-Diethyl-2.4-dinitro-6-(trifluoromethyl)-1,3-phenylenediamine
Dinitrophenol	
Dinoseb	2-(sec-Butyl)-4,6-dinitrophenol
Dioxathion	2,3-*p*-Dioxanedithiol *S,S*-bis(*O,O*-diethyl phosphorodithioate)
Diquat	6,7-Dihydrodipyrido (1,2-a:2′,1′-c)pyrazinediium dibromide, monohydrate
Disulfoton	*O,O*-Diethyl *S*-(2-[ethylthio]ethyl)phosphorodithioate
Disyston®	See disulfoton
Diuron	3-(3,4-Dichlorophenyl)-1,1-dimethylurea
DNP	Dinitrophenol
2-(2,4-DP)	2-(2,4-Dichlorophenoxy) propionic acid
Dursban®	See chlorpyrifos
Dylox®	See trichlorfon
Dyrene®	See anilazine
EDP	Ethyl diphenyl phosphate
EHDP	2-Ethylhexyl diphenyl phosphate
Enchlor®	See polychlorinated biphenyls
Endosulfan	6,7,8,9,10,10-Hexachloro-1,5,5a,6,9,9a-hexahydro-6,9-methano-2,4,3-benzodioxathiepin-3-oxide
Endothall	7-Oxabicyclo[2.2.1] heptane-2,3-dicarboxylic acid
Endrin	1,2,3,4,10,10-Hexachloro-6,7-epoxy-1,4,4a,5,6,7,8,8a-octahydro-1,4-endo, endo-5, 8-dimethanonaphthalene
EPN	*O*-Ethyl-*O*-(*p*-nitrophenyl) phenyl phosphonothioate
S-Ethyl *N,N*′-dipropyl thiocarbamate	
Ethyl parathion	See parathion ethyl
Fenitrothion	*O,O*-Dimethyl *O*-(4-nitro-*m*-tolyl) phosphorothioate
Fenoprop	2-(2,4,5-Trichlorophenoxy) propionic acid
Fenpropanate	*α*-Cyano-3-phenoxybenzyl-2,2,3,3-tetramethylcyclopropanecarboxylate

Fenvalerate	α-Cyano-3-phenoxybenzyl 2(4-chlorophenyl)-3-methyl butyrate
Fluridone	1-Methyl-3-phenyl-5-(3-trifluoromethylphenyl-4 (1H)-pyridinone
Furadan®	See carbofuran
GD-174	2-(Digeranylamino)ethanol
Glyphosate	N-(Phosphonomethyl) glycine
Glyphosate	Isopropyl amine salt
Guthion®	See azinophos methyl
HBB	Hexabromobenzene
HCB	See hexachlorobenzene
HCH	See hexachlorocyclohexane
HCP	Hexachlorophene
Heptachlor	1,4,5,6,7,8,8-Heptachloro-3a,4,7,7a-tetrahydro-4,7-methanoindene
Heptachlor epoxide	
Heptachlorobornane	
Hexachlorobenzene	
Hexachlorobiphenyl	See polychlorinated biphenyls
Hexachlorocyclohexane	$\alpha,\beta,\gamma,\delta,\varepsilon$ Isomers of 1,2,3,4,5,6-hexachlorocyclohexane; formerly called benzene hexachloride or BHC
Hydrothal and Hydrothal 191	See endothall
Isodrin	
Isopropyl PCBs	Isopropyl polychlorinated biphenyls
Kelevan	(Ethyl)-(5-hydroxyl-1,2,3,4,6,7,8,9,10,10-dicachloro-pentacyclo- $5.3.0.0^{26}.0^{39}.0^{48}$ decyl) levulinate
Kepone®	See chlordecone
Knox Out® 2FM	Formulation of diazinon; see diazinon
Leptophos	O-(4-Bromo-2,5-dichlorophenyl) O-methylphenylphosphonothioate
Lindane	γ-Isomer of hexachlorocyclohexane
Malathion	O,O-dimethyl S-(1,2-dicarbethoxyethyl) phosphorodithioate
MCPA	4-Chloro-2-methylphenoxy acetic acid
Merphos	Tributyl phosphorotrithioite
Metasystox®	See demeton-S-methyl
Methamidophos	See methazole
Methazole	2-(3,4-Dichlorophenyl)-4-methyl-1,2,4-oxadiazolidine-3,5-dione
Methoprene	Isopropyl (2E,4E)-11-methoxy-3,7,11-trimethyl-2,4-dodecadienoate
Methoxychlor	2,2-bis(p-Methoxyphenyl)-1,1-trichloroethane
Methylmercury	
Methyl parathion	See parathion methyl
Mevinphos	Dimethyl phosphate of methyl-3-hydroxy-cis-crotonate
Mexacarbate	4-(Dimethylamino)-3,5-xylyl methylcarbamate
Mirex	Dodecachloroctahydro-1,3,4-methano-2H-cyclobuta(cd)pentalene
Molinate	S-Ethyl hexahydro-1 H-azepine-1-carbothioate
MSMA	Monosodium methanearsonate

Nabam	Disodium ethylene bis-dithiocarbamate
Naled	1,2-Dibromo-2,2-dichloroethyl dimethyl phosphate
Nitrofen	2,4-Dichlorophenyl-*p*-nitrophenyl ether
Nitrosalicylinides	3'-Chloro-3-nitrosalicylanilide
Nonachlor	
Nonylphenol	
NPDPP	Nonylphenyl diphenyl phosphate
Octachlorostyrene	
Oxychlordane	
Parathion ethyl	*O,O*-Diethyl *O-p*-nitrophenyl phosphorothioate
Parathion methyl	*O,O*-Dimethyl-*O-p*-nitrophenyl phosphorothioate
PBBs	See polybrominated biphenyls
PCA	See pyrazone
PCBs	See polychlorinated biphenyls
PCDF	Polychlorinated dibenzofuran
PCP	See pentachlorophenol
Penncap M®	Formulation of methyl parathion; see parathion methyl
Pentachlorobenzene	
Pentachlorobiphenyl	See polychlorinated biphenyls
Permethrin	3-(Phenoxyphenyl)methyl(I)-*cis, trans*-3-(2,2-dichloro-ethenyl)-2,2-dimethylcyclopropanecarboxylate
Phenochlor	See polychlorinated biphenyls
Phorate	*O,O*-Diethyl *S* ([ethylthio]methyl) phosphorodithioate
Phosalone	*O,O*-Diethyl *S*-(6-chloro-3-[mercaptomethyl]-2-benzox-azolinone) phosphorodithioate
Phosmet	*N*-(Mercaptomethyl) phthalimide *S*-(*O,O*-dimethyl-phosphorodithioate)
Phosphamidon	2-Chloro-*N,N*-diethyl-3-(dimethyloxyphosphinyloxy) crotonamide
Phosphoric acid triesters	See pydraul
Phthalate esters	See phthalic acid esters
Photoaldrin	
Photochlordane	
Photodieldrin	
Photoheptachlor	
Picloram	4-Amino-3,5,6-trichloropicolinic acid
Piscicide GD-174	See GD-174
Polychlorinated biphenyls	
Polychlorinated terphenyls	Aroclors® with 54% chlorine
Profenofos	*O*-(4-Bromo-2-chlorophenyl) *O*-ethyl *S*-propyl phos-phorothioate
Propanil	3',4'-Dichlorophenylpropionanilide
Propoxur	*O*-Isopropoxyphenyl *N*-methylcarbamate
Pydraul	Mixture of tri-aryl phosphate esters
Pyralene	See polychlorinated biphenyls
Pyrazone	5-Amino-4-chloro-2-phenyl-3 (2 *H*)-Pyridazinone
Pyrethrum	Mixture of natural pyrethrins
Quinalphos	*O,O*-Diethyl-*O*-(2-quinoxalinyl)-phosphorothioate
Reglone®	See diquat
Reldan®	*O,O*-Dimethyl *O*-(3,5,6-trichloro-2-pyridyl) phospho-rothioate

Resmethrin	(5-Benzyl-3-furyl)methyl 2,2-dimethyl-3-(2-methyl propenyl) cyclopropanecarboxylate
Rogor®	See dimethoate
Ronnel	*O,O*-Dimethyl *O*-(2,4,5-trichlorophenyl) phosphorothioate
Rotenone	1,2,12,12a-Tetrahydro-2-isopropenyl-8-9-dimethyoxy(1)benzopyrano (3,4-b)furo (2,3-b) (1)benzopyran-6(6 AH)-one
Round Up®	See glyphosate
Rovral®	3-(3,5-Dichlorophenyl)-*N*-(1-methylethyl)-2,4-dioxo-1-imidazoline carbonamide
Santotherm®	See polychlorinated biphenyls
SBP-1382	See resmethrin
Sesamex	5-[{1-(2-[2-Ethoxyethoxy)ethoxy)ethoxy}]-1,3-benzodioxole
Silvex®	See 2,4,5-T
Simazine	2-Chloro-4,6-bis(ethylamino)-s-triazine
SKF-525A	
Sodium arsenite	
Strobane®	Polychlorinates of camphene, pinene, and related terpenes
Systox®	See demeton
2,4,5-T	2,4,5-Trichlorophenoxy acetic acid
TBTO	See bis(tributyltin oxide)
TCA	See trichloroacetic acid
TCDD	See tetrachloro dibenzo dioxin
TCDF	See tetrachloro dibenzo furan
TCP	Tricresyl phosphate
TDCPP	Tris (1,3-dichloroisopropyl) phosphate
Telodrin	
Temephos	*O,O,O′ O′*-Tetramethyl *O,O′*-(thiodi-*p*-phenylene)diphosphorothioate
TEPA	Tris (1-aziridinyl) phosphine oxide
TEPP	Tetraethyl diphosphate; tetraethyl pyrophosphate
Terbutryn	2-(tert-Butylamino)-4-(ethylamino)-6-(methylthio)-s-triazine
Tetrachlorobiphenyl	See polychlorinated biphenyls
Tetradifon	4-Chlorophenyl 2,4,5-trichlorophenyl sulfone; (2,4,5,4′-tetrachlorodiphenyl sulfone)
TFM	3-Trifluoromethyl-4-nitrophenol, sodium salt
Therminal	See polychlorinated biphenyls
Thidiazuron	*N*-Phenyl-*N′*-(1,2,3-thiadiazol-5-yl)urea
Thimet®	See phorate
Thiobencarb	S-(4-Chlorobenzyl)*N,N*-diethylthiocarbamate
Toxaphene	Chlorinated camphene (67 to 69% chlorine) mixture
TPP	Triphenyl phosphate
Trichlorfon	Dimethyl(2,2,2-trichloro-1-hydroxyethyl) phosphonate
Trichloro biphenyl	See polychlorinated biphenyls
2,4,6-Trichlorophenyl	*p*-Nitrophenyl ether
Trichlorophenol	

Trichloropyridinol
Trifluralin a,a,a-Trifluoro-2,6-dinitro-N,N-dipropoyl-p-toludine
Tris phosphate Tris (2,4-di-ter-butylphenyl)phosphate
Trithion® See carbophenothion
Zolone® See phosalone

Appendix II

COMMON NAMES OF FISH MENTIONED IN THE TEXT AND THEIR SCIENTIFIC NAMES

African lakefish	*Tilapia* sp.
Alewife	*Alosa pseudoharengus* (Wilson)
American shad	*Alosa sapidissima* (Wilson)
American smelt	*Osmerus mordax*
Arctic char	*Salvelinus alpinus* (Linn.)
Atlantic cod	*Gadus* sp.
Atlantic croakers	*Micropogon* sp.
Atlantic salmon	*Salmo salar* Linn
Atlantic silverside	*Menidia menidia*
Baltic flounder	*Platichthys flesus*
Barbus sp.	
Bass	See white bass and yellow bass
Bigmouth buffalo	*Ictiobus cyprinellus* (Valenciennes)
Black bullheads	*Ictalurus melas* (Rafinesque)
Black crappie	*Pomoxis nigromaculatus* (LeSueur)
Black surfperch	*Embiotoca jacksoni*
Bluegills	*Lepomis macrochirus* Rafinesque
Bluntnose minnows	*Pimephales notatus* (Rafinesque)
Brook trout	*Salvelinus fontinalis* (Mitchill)
Brown bullheads	*Ictalurus nebulosus* (LeSueur)
Brown trout	*Salmo trutta*
Bullheads	See brown bullheads
Burbot	*Lota lota* (Linn.)
Carp	*Cyprinus carpio* Linn.
Channa	*Channa punctata*
Channel catfish	*Ictalurus punctatus* (Rafinesque)
Char	See Arctic char
Chinook salmon	*Oncorhynchus tschawytscha* (Walbaum)
Chub	*Coregonus alpenae*
Cirrhinus mrigala	
Cisco	*Coregonus artedii* LeSueur
Cod	*Gadus morrhua*
Coho salmon	*Oncorhynchus kisutch* (Walbaum)
Common shiner	*Notropis cornutus* (Mitchill)
Convict cichlid	*Cyclosoma nigrofasciatum*
Creek chub	*Semotilus atromaculatus* (Mitchill)
Crucian carp	*Carassius carassius*
Cutthroat trout	*Salmo clarki* Richardson
Cynoglossus	
Dogfish	*Squalus* sp.
English sole	*Parophrys vetulus*
Fathead minnow	*Pimephales promelas*
Flagfish	*Jordanella floridae*
Flounder	*Platichthys* sp.
Gizzard shad	*Dorosoma crepedianum* (LeSueur)
Glyphocephalus	
Gobi fish	*Gobius* sp.

Goldenshiners	*Notemigonus crysoleucas* (Mitchill)
Goldfish	*Carassius* sp.
Goldorfe	*Leusescus ides*
Green sunfish	*Lepomis cyanellus* Rafinesque
Guppy	*Poecilia* sp.
Harlequin fish	*Rosbora heteromorpha*
Herring	*Clupea harengus*
Johnny darter	*Etheostoma nigrum* Rafinesque
Killifish	*Oryzeas latipes*
King salmon	*Oncorhynchus tshawytscha*
Labeo rohita	
Landlocked salmon	*Salmo salar* Linn.
Lake herring	See cisco
Lake trout	*Salvelinus namaycush* (Walbaum)
Lamprey	*Petromyzon* sp.
Largemouth bass	*Micropterus salmoides* (Lacépède)
Leucaspius dileneatus	
Limnothrissa	
Longear sunfish	*Lepomis megalotis*
Longnose killifish	*Fundulus similis*
Longnose sucker	*Catostomus catostomus*
Medaka	*Oryzias latipes*
Menhaden	*Brevoortia* sp.
Minnow	*Phoxinus phoxinus*
Mosquitofish	*Gambusia affinis* (Baird and Girard)
Motsugo fish	*Pseudorasbora parva*
Mullet	*Mugil* sp.
Mystus cavasius	
Northern anchovy	*Engraulis modax*
Northern puffers	*Sphaeroides maculatus*
Paddlefish	*Polyodon spathula* (Walbaum)
Perch	*Perca fluviatilis*
Pike	*Esox lucius* Linn.
Pinfish	*Lagodon rhomboides*
Pumpkin seed	*Lepomis gibbosus* (Linn.)
Rainbow trout	*Salmo gairdneri* Richardson
Redear sunfish	*Lepomis microlophus* (Gunther)
Redhorse suckers	*Moxostoma macrolepidotum* (LeSueur)
Reticulate sculpins	*Cottus perplexus*
Riogrande perch	*Cichlasoma cyanoguttatum*
Rivercarp sucker	*Carpiodes carpio*
Roach	*Rutilus rutilus*
Sailfin mollies	*Poecilia latipinna*
Salmon	*Salmo* sp.
Sand trout	
Sardine	*Sardina* sp.
Sea trout	*Cynoscion* sp.
Sheepshead minnow	*Cyprinodon variegatus*
Shiners	See golden shiners
Smallmouth bass	*Micropterus dolomieui* Lacépède
Sockeye salmon	*Oncorhynchus nerka* (Walbaum)
Speckled trout	

Spot	*Leiostomus xanthurus*
Statothrissa	
Steelhead trout	Sea running phase of the rainbow trout *Salmo gairdneri*
Stickleback	*Gasterosteus* sp.
Striped bass	*Morone saxatilis*
Suckers	*Catostomus* sp.
Swordfish	*Xiphias gladias*
Therapon jarbua	
Thorny skate	*Raja radiata*
Tilapia	
Topmouth gudgeon	*Pseudorasbora parva*
Varichorhinus	
Wall eye	*Stizostedion vitreum vitreum* (Mitchill)
White bass	*Roccus chrysops* (Rafinesque)
White crappie	*Pomoxis annularis* Rafinesque
White fish	*Coregonus clupeaformis* (Mitchill)
White perch	*Roccus americanus* (Gmelin)
White suckers	*Catostomus commersoni* (Lacépède)
Winter flounder	*Pseudopleuronectes americanus*
Xiphias gladias	
Yellow bass	*Roccus mississippiensis* (Jordan and Eigenmarine)
Yellow bullheads	*Ictalurus natalis* (LeSueur)
Yellow perch	*Perca flavescens* (Mitchill)
Zebra danio	*Brachydanio rario*

INDEX

A

Printed and bound by CPI Group (UK) Ltd, Croydon, CR0 4YY

22/10/2024

01777630-0017